Creating WiMAX RFPs

Anthony Klinkert

Althos Publishing
Fuquay-Varina, NC 27526 USA
Telephone: 1-800-227-9681
Fax: 1-919-557-2261
email: info@althos.com
web: www.Althos.com

Althos

All rights reserved. No part of this book may be reproduced or transmitted in any form or by any means, electronic or mechanical, including photocopying recording or by any information storage and retrieval system without written permission from the authors and publisher, except for the inclusion of brief quotations in a review.

Copyright © 2008 By Althos Publishing
First Printing

Printed and Bound by Lightning Source, TN.

"WiMAX," "Mobile WiMAX," "Fixed WiMAX," "WiMAX Forum," "WiMAX Certified," "WiMAX Forum Certified," the WiMAX Forum logo and the WiMAX Forum Certified logo are trademarks of the WiMAX Forum.

> Every effort has been made to make this manual as complete and as accurate as possible. However, there may be mistakes both typographical and in content. Therefore, this text should be used only as a general guide and not as the ultimate source of information. Furthermore, this manual contains information on telecommunications accurate only up to the printing date. The purpose of this manual to educate. The authors and Althos Publishing shall have neither liability nor responsibility to any person or entity with respect to any loss or damage caused, or alleged to be caused, directly or indirectly by the information contained in this book.

International Standard Book Number: 1-932813-95-0

Acknowledgements

There were many gifted people who gave their technical and emotional support to me in a way that helped me create this book. In many cases, published sources were not available on this subject area. Experts from academia, manufacturers, system integrators, trade associations, prior employers and other wireless related companies, associates, and colleagues gave their personal and precious time to help me, and for this I sincerely thank and respect them.

All these professionals provided insight and inspiration at some point along the journey. They include, from SMU, Dr. Richard Barr, Dr. Joakim Kalvenes, and Dr. Eli Olinick, from the WCA, Andrew Krieg, from EDX, Jennifer Duncan, from PBM, Shiva Kavi, and Ted Sumner, from the WiMAX Forum, Ron Resnick, and Dr. Mo Shakouri, from CelPlan, Leonhard Korowajczuk, and Oscar Miranda, from IBM Stephen Currie, Philip Mullins, Christopher Douglass, Charlie Arteaga, and Lisa Schmidt, from PCN Pros, Bruce Blais, from Sprint Nextel, Harry Perlow, from Telecom Ventures, Raj Singh, from TeleworX, Jose Rodriguez, from Blue Moon Solutions, Marty Hale, from Comcast, Kerry McKelvey, from Google, Vint Cerf, from LBA Group , Lawrence Behr, from Memorylink, Thomas Freeburg and from WFI, Bob Shapiro. My thanks again go to you for contributing in significant ways.

 My deepest gratitude and dedication of this book goes to my wife Susan and children Lauren and Jake for their understanding and encouragement during the writing of this work. I also offer a heartfelt thank you to my parents Lawrence and Sofia for the can do sprit they established in our family.

Special thanks finally go to the people who assisted with the production of this book including Carolyn Luck (Director), Michele Chandler (Associate Editor), Geovanny Solera (Researcher), and Lawrence Harte (Editor and Publisher).

About the Author

Anthony (Tony) Klinkert has established a distinguished career as an executive in corporate management, as a senior consultant with blue chip and entrepreneurial consultancies, and in engineering posts in such leading corporations as IBM, MCI, GE, TI, and Motorola. He successfully established entrepreneurial ventures, including Klinkert and Associates, Inc., a licensed professional engineering services corporation and consultancy in the emerging broadband wireless sector.

Tony has approximately three decades of experience in radio and wireless network planning, modeling, design, deployment and operations research applied to wireless design. His experience includes international nationwide cellular system design, the planning and technology strategy for the nation's largest business-class pre-WiMax broadband wireless network, technology strategy for the world's largest advanced messaging network, and over 300 projects for wide area land-mobile and short-haul point-to-point systems. As CEO and principal consultant at Klinkert & Associates, noteworthy engagements include advising Fortune 100 corporations and entrepreneurs on, a nationwide 600-location WLAN strategy and design, a wide area outdoor broadband wireless network, and a nationwide broadband wireless program. His specialty is providing advanced management science modeling and optimization of wireless technology for entrepreneurs, executives and officials establishing technology-based revenue, cost, income and discounted cash flow models for business cases covering major wireless network initiatives.

Tony is also an award winning industry leader, having won the broadband wireless industry association's highest award, The WCA Golden Eagle, for "technical vision and industry service". He is a frequent public speaker, having represented a global carrier's technology strategy at conventions, web casts, and in the trade press.

He received his BSEE from the University of Texas at Austin, and both an MSEE, and an MS-Telecommunications from Southern Methodist University. His Doctor of Engineering in Engineering Management focuses on the application of advanced operations research mathematical programming techniques to the optimization of wireless design and deployment business models.

Tony is a federally licensed radio operator, a state licensed professional engineer, has served on local government telecommunications policy committees and is a frequent contributor to standards bodies in emerging technology areas of the wireless telecommunications sector.

Tony is currently assisting IBM Global Technology Services' National Wireless and Mobility consulting practice. There he consults to C-Level executives and officials responsible for wireless and mobility initiatives.

Table of Contents

INTRODUCTION 1

SOURCING 2

    NON-COMPETITIVE SOURCING 3
    COMPETITIVE SOURCING 4
    PILOT OR TRIAL 4
    IMPLEMENTATION, OPERATION, REFRESH AND END OF LIFE 4

REQUEST FOR PROPOSALS 5

    REQUEST FOR QUALIFICATION (RFQ) 7
    REQUEST FOR INFORMATION (RFI) 8
    SYSTEM SOLUTIONS 8
    REQUIREMENTS STATEMENTS 9

USING RFPS 11

    KEY RFP OBJECTIVES 11
        Optimal Solutions *12*
        Vendor Selection *12*
        Successful Service Launch *12*
    DECISION CRITERIA 13

THE NEED FOR WIMAX 14

    BROADBAND PRIVATE DATA NETWORK 14
    BROADBAND PUBLIC NETWORKS AND INTERNET ACCESS 14

DIGITAL TELEPHONE AND VOIP	15
DIGITAL TELEVISION AND VIDEO	16

AN OVERVIEW OF WIMAX ... 19

WIRELESS	19
BROADBAND WIRELESS	20
TECHNOLOGY STANDARDS	20
CERTIFICATIONS	21
SPECTRUM	21
RADIO SYSTEM	21
RADIO AND DATA COMMUNICATIONS	22
NETWORKS AND INFORMATION TECHNOLOGY	22
WIMAX STANDARDS AND CERTIFICATION	23
WIMAX NETWORK	24
Fixed Service	*26*
Nomadic and Portable Service	*27*
Mobile Service	*27*
WIMAX SPECTRUM	27
Spectrum Due-Diligence	*28*
License Exempt Spectrum	*28*
Spectrum Details for a particular region	*29*
Spectrum Available for WiMAX	*30*
WIMAX CERTIFICATION	31
WiMAX Profiles	*32*
Profiles Related to WiFi	*33*
Certification Not Wanted	*34*

WIMAX REQUIREMENTS ... 34

WIMAX BUSINESS REQUIREMENTS ... 35

APPLICATIONS, SERVICES AND USE CASES REQUIREMENTS	35
USE-CASES	36
SERVICE AREA & COVERAGE REQUIREMENTS	36
MARKETING AND SELLING REQUIREMENTS	37

FINANCE AND ACCOUNTING REQUIREMENTS 37
BUSINESS TO TECHNICAL REQUIREMENTS 37

WIMAX TECHNICAL REQUIREMENTS **38**
SERVICE AND COVERAGE REQUIREMENTS 39
Extent of Coverage *40*
Range of Coverage *41*
Bandwidth and Throughput *41*
Types of Coverage Levels and Location Availability *42*
Cost of Coverage *43*
Outdoor Coverage *43*
Indoor Coverage *43*
Link Budget *44*
Site Surveys *46*
DATA NETWORK REQUIREMENTS 46
INTERNET ACCESS REQUIREMENTS 47
VOICE AND TELEPHONY REQUIREMENTS 49
VIDEO AND TELEVISION REQUIREMENTS 49
ACCESS DEVICES REQUIREMENTS 50
Premises Distribution Equipment *51*
DISTRIBUTION NETWORK REQUIREMENTS 52
Network Architecture *52*
Protocols *53*
Data Capacity *53*
Criticality *53*
Scalability *53*
Servers *53*
TESTING REQUIREMENTS 54
Field Tools *54*
System Diagnostics *54*
Self-Diagnostics *55*
Acceptance Testing and Commissioning *55*
AUTHENTICATION, AUTHORIZATION, AND ACCOUNTING (AAA) 55
CONDITIONAL ACCESS SYSTEM (CAS) 56

SYSTEM ADMINISTRATION 56
   Service Provisioning *56*
   Remote Administration *57*
CUSTOMER CARE 57
BILLING SYSTEMS 58
RELIABILITY, AVAILABILITY AND MAINTAINABILITY (RAM) 60
SINGLE POINT OF FAILURE, REDUNDANCY, AND BATTERY BACKUP . . 60
DISASTER RECOVERY 61

THE RFP DEVELOPMENT PROCESS 62

QUALIFICATIONS 64
   Financial Qualifications *64*
   Technical Qualifications *64*
   Operating Information *64*
   Licenses *65*
   Certifications *65*
SUPPORTING VENDORS 65
EXISTING SYSTEM EVALUATION 65
   Distribution Plant *66*
   Access Systems *66*
   Back Office Systems *66*
NEEDS ASSESSMENT 66
   Advisory Committee *67*
   Visioning Session *67*
ISSUING, APPROVAL AND CONTRACT AUTHORITY 69
   Public Notice *69*
   Contracting Authority *70*
RFP PROJECT DATES 70
   Conferences *70*
   Clarification Requests *70*
   Clarification Responses *71*
   Pre-Response Conferences *71*
   Site Review *71*
   RFP Release Date *71*
   Closing RFP Submission Date *71*

    RFP Award Date 72
    Contract Negotiation Date 72
    Final Signing 72
  IMPLEMENTATION PLAN 73
    Initial Operational Capability (IOC) Date 74
    Alpha Testing Date 74
    Field Trial Date (Beta Testing) 74
    Full Operational Capability (FOC) 75
    System Cutover 75
    Acceptance Testing 75
  RFP RESPONSE PROCEDURE 75
  RFP ISSUANCE 76
  PROSPECTIVE SUPPLIERS 76
  RFP RESPONSE REQUIREMENTS 79
  RFP APPROVAL PROCESS 80
  PRE-PROPOSAL REVIEW 80
  RELEASING RFPS 80

RFP RESPONSES 80

  RFP RESPONSE REVIEW 81
  CLARIFICATION QUESTIONS 81
  CLARIFICATION RESPONSE REVIEW 81
  ISSUING CLARIFICATION UPDATES 82

RESPONSE EVALUATION 82

  EVALUATION CRITERIA 82
  DECISION MATRIX 84
  CRITERIA WEIGHTING 84
  SCORING 84
  WINNER SELECTION 85
  SHORT-LISTED VENDOR MEETINGS 87
  NOTIFICATION OF AWARD 87
  NOTIFYING OF REJECTION 87
  PROTEST OF AWARD 88

RFP CONTENTS 88
  RFP FOR A WiMAX NETWORK TEMPLATE 88
  SUMMARY OF NEEDS AND PURPOSE STATEMENT 91
  GENERAL PROCEDURES 91
    RFP Submission Procedure *91*
    RFP Transmittal Process *92*
  ISSUING COMPANY INFORMATION 92
    RFP Requirements Summary *92*
    Company Background *92*
    Existing Systems and Services *93*
    RFP Objectives *93*
    Scope of Work *93*
    Instruction to Respondents *93*
    RFP Distribution *93*
    Bidders Conference *93*
    RFPs Clarification *93*
    Response Requirements *94*
  RESPONDENT INFORMATION 94
    Respondent Company Information Request *94*
  WiMAX NETWORK REQUIREMENTS 94
    Business Requirement *94*
    Technical Requirements *94*
  IMPLEMENTATION SCHEDULE 95
    Testing Requirements *96*
  TRAINING 98
  ACCEPTANCE 98
  PROCUREMENT TERMS AND CONDITIONS 99
    Terms and Conditions of Proposal *99*
    Change Orders *100*
    Compliance *100*
    Nonresponsive Proposal *100*
    Performance Guarantees *101*
    Proprietary Information *101*
    Warranty and Guarantees *101*

Regulatory Compliance *102*
Liability *102*
Right to Reject *102*
PRICING AND FINANCING OPTIONS 102
Pricing Options *102*
Elements in the Business Case *103*
Equipment Pricing *103*
Support Services *103*
Financing Terms *104*
Submission of Financial Information *104*

AFTERWORD **105**
References: *105*

APPENDIX 1 - ACRONYMS **107**

APPENDIX 2 - GLOSSARY **111**

APPENDIX 3 - WIMAX RESOURCES **129**

APPENDIX 4 - WIMAX RFP TEMPLATE **133**

INDEX **137**

Introduction

This book provides the reader with information on how to develop a WiMAX Request for Proposal (RFP). This book is an excellent guide for executives, senior staff and decision influencers that need to understand the elements that go into creating an RFP for a complex wireless network. Focusing more on business process than engineering, this book is written in a fact-based style, without the frills that surround much of the marketing discussion on WiMAX available today. We also maintain in this book the publisher's style of "knowledge elements" containing "just the facts" in a quick, get to the point, fashion. The inclusion of a detailed table of contents, complete index, web reference, glossary, and acronyms list provide the ability to quickly find information. We strive for brevity, clarity, and accessibility of key knowledge elements, to make the book a quick read for those developing a WiMAX RFP, including C-level executives, legal, procurement and technical staff, and a ready reference for those needing a refresher on any specific element of the process. For more information on the subject, including a readers' forum, visit www.CreatingWiMAXrfps.com

This book provides general information on RFPs, including its comparison with Request for Information (RFI), Request for Qualification (RFQ) and other types of formal interaction with vendors. It develops the key objectives, decision criteria, development process, qualification and selection of vendors, the needs assessment, implementation plans, issuing and approval authority, and the approval process. It also describes the question and clarification process and the evaluation process consisting of the decision matrix, the criteria for computing the winning respondent and how to handle notification of award and rejection. The book provides a complete overview of the process for developing WiMAX system RFPs generically.

Most importantly, the book delves into the WiMAX technical requirements that are crucial to understanding a vendor's solution and judging its merits against the needs of the RFP issuer. This book covers the development of the technical requirements for the numerous factors crucial for approaching an

RFP for a WiMAX solution. These factors include technical requirements for the applications, services, service area, marketing, selling, financing, accounting, and general and administrative aspects of a WiMAX network. It also includes the technical requirements for the data network, voice network, video network, core platforms, distribution network, and access devices. Key areas covered also include testing and acceptance, conditional access, system administration, customer care, billing, and disaster recovery.

Lastly, a sample WiMAX RFP is available on the author's web site that will offer the reader a quick start and guide for developing a specific WiMAX RFP for a distinctive network operator, and in particular implementation scenario.

Sourcing

When an organization embarks on procurement of a major asset, such as a WiMAX solution, the organization typically appoints the Procurement department to establish the proper procedures. Procurement officers of an organization describe the acquisition of new assets as *sourcing*. The sourcing process is shown as part of the complete life of a system as shown in figure 1. As part of the sourcing process, the organization must make a sourcing decision, namely, if the network will be sourced in a competitive or non-competitive fashion.

Non-competitive sourcing

If the network is sourced in a non-competitive fashion, as a sole-source to a preferred vendor, the purchaser contacts the vendor, discusses purchaser needs, then the vendor completes a design and proposal for the purchaser. After some negotiations, taking into account any new information or changes, a contract for purchase can ensue. A non-competitive sourcing process is best used when a strategic relationship with one vendor exists and time is of the essence (above cost and performance).

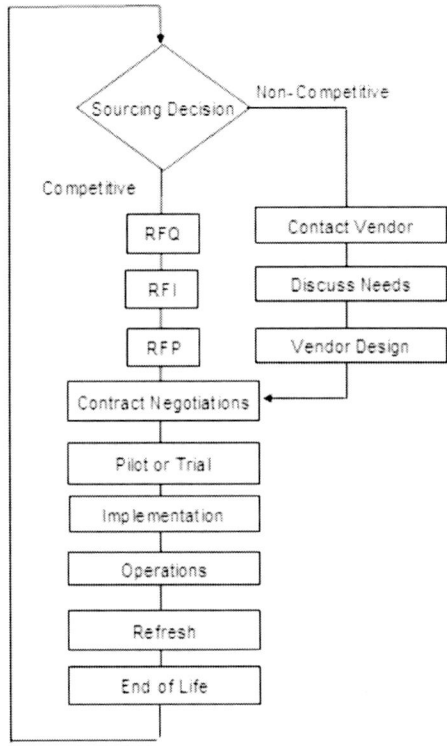

Figure 1: System Life Cycle

Competitive Sourcing

If the purchaser decides to source the network competitively, an RFQ, RFI and RFP process can ensue. After that process, and further negotiations, taking into account any new information or changes, a contract for purchase can ensue. A competitive sourcing process is best used when cost of procurement is important and performance is critical. A competitive sourcing process takes longer than a non-competitive sourcing process, so it should be undertaken when priority is placed upon performance and cost versus time.

Pilot or Trial

After contract negotiations, a pilot or trial of the solution is possible whereby the purchaser tests a small subset of the solution to learn details regarding the technology and implementation. The trial is extended to specific users, potentially aspects of sales and marketing. If the network is sourced as a competitive bid, a *bake off* can take place, whereby two vendors are offered the opportunity to perform a pilot or trial and the performance, cost and implementation execution can be compared by the purchaser.

Implementation, Operation, Refresh and End of Life

Once contract negotiations and any pilot or trial are complete, an implementation or deployment of the main WiMAX network can begin. The deployment of the network requires good *program management* to ensure that the purchaser's requirements are met in the calendar (and labor resource) time required, as well as within the budget established.

The purchaser will accept the system if the vendor meets all the technical specifications. The success of the test acceptance, therefore, is highly dependent upon having testable specifications by which to judge compliance with contact performance agreements and network operator requirements.

Once the system is accepted, operations begin. Business and operational support systems are important since they ensure efficient network functioning in terms of monitoring performance, any required billing, and receiving alarm indications for fault conditions. During operations, network maintenance will take place. It is imperative that the vendor support the operator with services and capabilities to efficiently maintain the network, including clearing faults, initiating updates, maintaining spares and repair parts and performing remote configuration.

As a network matures, there will be a need for *technology refresh,* which refers to the ability of the vendor to provide a major upgrade of functionality or capability. Over a long planning horizon, possibly 10 years, this may occur on a three to five year period and contractual provisions must be anticipated and in place to handle this cost-efficiently.

Lastly, a network will ultimately reach end of life when the vendor no longer provides support in terms of repair or spare parts. At that point, a new network is required and the sourcing process begins again.

Request for Proposals

A Request for Proposal (RFP) is a formal statement of technical requirements that clearly constitutes a company's need for the evaluation, recommendation and bid proposal of systems, networks, services or products from vendors. An RFP for a WiMAX system covers the communication requirements that will be satisfied by systems that use IP protocols such as Internet access, Voice over IP (VoIP) for telephone services, and video services including television and video conferencing. With sufficient spectrum, these services can be delivered and compete favorably against traditional telephone company dial-up or DSL service, or Cable company Internet access for data, DSL based TelcoTV, cable based IP CATV and other types of access systems and services.

RFPs are used for procuring systems and services for a company's business and operational requirements. RFPs that are used for WIMAX systems define the desired services that a company wants to provide, along with other key requirements such as integration into existing systems, management control and expansibility. The company who creates an RFP is called an RFP issuer and the people or companies that respond to an RFP are called RFP respondents or bidders.

RFPs typically include general information along with detailed product information, terms & conditions, and pricing and financing requirements that are associated with a project or group of projects that a company is initiating to allow them to offer new or improved products and services.

The general information section of an RFP provides information about the requesting company and solicits information about the responding companies as to their capabilities and abilities to meet the need. The technical requirement section covers the specific requirements that the requesting company cites in order to ensure that its needs are satisfactorily met. The terms and conditions section specifies the terms in which the requesting company has asked that the services and equipment be provided and the conditions under which it should be provided. The pricing and financing options section specifies the pricing and financing options the requesting company is seeking and an opportunity for the responding company to cite the pricing and financing options that they are willing to supply.

Figure 2 shows an outline of key elements of an RFP. This diagram shows that an RFP typically starts with a general overview of what is being requested. Several pages that detail key requirements such as product types, services and support needs usually follow this. The RFP also includes the terms & conditions that define how the vendor should respond such as who to respond to (contact information), in what format to respond (printed and/or electronic file formats), and key response submission dates.

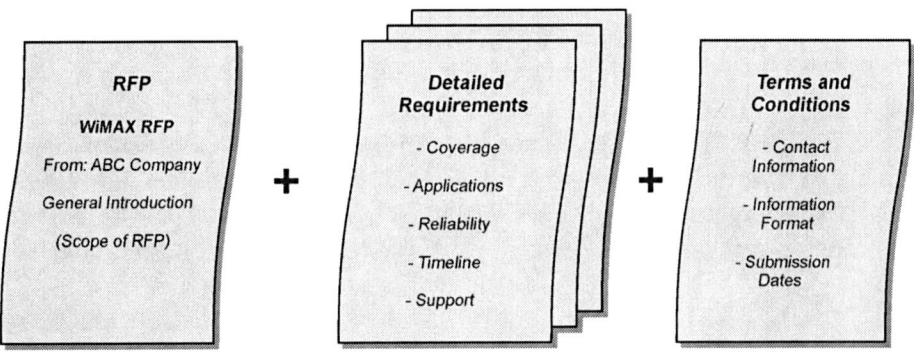

Figure 2: Key RFP Elements

Request for Qualification (RFQ)

Prior to issuing an RFP, a company will often issue a Request for Qualification (RFQ). An RFQ is a document that an issuer creates and distributes to companies that are believed to be interested in selling equipment, a total solution or services to the issuer. RFQs may ask companies to provide information on their financial, technical, operational experience, licenses and certifications. The RFQ also asks the respondents to provide background information on their company and capabilities regarding an issuer's opportunity. Lastly, the RFQ can be used to identify a complete list of consultants, resellers, manufacturers, system integrators, or other vendors interested in the opportunity. It is best to start the RFP process by issuing an RFQ.

Request for Information (RFI)

A request for information is a formal statement of information needs that may be used to determine more information regarding the feasibility of products, services or systems. An RFI may be issued after an RFQ, and prior to an RFP to help determine more specifically what to include in an RFP.

An RFI can be a document issued to companies, with a written response required, or it can be a response in the form of a structured meeting that provides answers to a list of specific questions of interest to the issuer (or both). This ensures an "apples to apples" comparison of vendor information.

An RFI can be used to obtain a high level response from the respondents regarding their approach or outline to the issuer's opportunity. It can also be used to finalize a short-list of vendors for a subsequent RFP from a broader list of potential vendors.

The RFI allows a first look at a broad high level response of performance, coverage, architecture, and schedule topics.

System Solutions

A system solution is an engineered design accompanied by a price for a total package of specific application(s) and the engineering required to guarantee the required quality and performance from the application platforms in the core, through the network, to the user's devices, or *"end to end."* This includes wireless coverage quality and extent of the service and applications, and lastly the implementation in a specific timeframe. One can see that a system solution is much more complex than a simple equipment bid.

An equipment bid is a price quote for a piece of equipment, or a list of equipment, sometimes called a bill of materials (BOM). A simple equipment bid is possible for a very small procurement, or simple expansions to existing networks. Occasionally, a very small order or an order for a commodity is referred to as a *request for quote (Rfq)*.

For equipment bids, a *reseller* of manufacturers' products should be able to provide a price quotation for the BOM, and may be able to provide an installation quotation, but likely may need to seek subcontractors for engineering if some engineering design is needed, or if extensive installation is needed. For system solutions, it is better to go directly to larger manufacturers who likely have the capability to quote a full wireless solution, including design, engineering and installation of the wireless network.

For larger and more complex network solutions, the best respondents are likely *system integrators (SI)*. For large systems, SIs can act as consultants and technical advisors to the issuer to ensure all aspects of a solution are designed, deployed and tested to the issuer's technical requirements. The SI can have total responsibility for a turn-key solution, or simply *program management* and responsibility to act as a third party "trusted technical advisor." Since SIs do not necessarily sell hardware or software, rather they sell *professional services*, they have the objective credibility to assist in vendor selection at the conclusion of an RFP process, and then to assess the quantity of a vendor's equipment specified. This precludes the situation of a dominant vendor that might over-engineer a solution (given a strong position against competitors) or a weak vendor that might under-engineer a solution (in order to win a competitive bid). Programs that use SIs are sometimes called Engineer, Furnish, and Install (EFI) or Engineer, Procure, and Construct (EPC) projects.

For the largest systems, SIs typically act as *prime contractors*, responsible for the entire *turn-key,* or *design-build*, solution in an environment that may have several manufacturers of products and services as diverse as radio, data networking, operational support systems, business support systems, Internet and VoIP gateway services, mass deployment, site acquisition, construction and programming content for IPTV.

Requirements Statements

Requirements statements are assertions of anticipated needs. Requirements statements may be definitive (descriptive) or probing (to promote the discovery of additional information).

A definitive requirement states a need of the issuer directly. The respondent is expected to comply with the requirement as stated. If it is a business or general requirement, the respondent must indicate compliance, and provide explanatory text to indicate understanding or clarification. If it is a technical requirement, the respondent should provide detailed technical specifications that can be tested and document how the respondent's design solution meets the intent of the technical requirement. For example, a definitive technical requirement may be: "The respondent shall provide a mean time between failures (MTBF) of 10 years ". The vendor's response should include testable technical specifications of the vendors' products and solutions.

A probing technical requirement states the high level intent of a requirement then requires the respondent to make the requirement specific to the vendor's equipment or solution. For example, "The respondent shall provide a highly reliable base station product with minimal failure throughout its lifecycle and provide a detailed description of the MTBF, method of calculation, and any accelerated life testing performed." In this example, a specific numerical technical requirement is not provided, rather the intent of the requirement is described, and the expectation is that the vendor should offer information specific to the vendor's solution and equipment. The vendor's response, after the vendor has designed the vendor-specific solution, should quantifiably establish a measurable technical requirement with technical specifications that meet the intent of the probing technical requirement of the issuer.

Using RFPs

RFP's can be used to help ensure that the optimal supplier selected will provide a solution that is well matched to the defined future needs of the company. The creation of an RFP is a process that involves internal analysis along with information that is required from other sources (vendors).

The RFP development process involves more than producing a document that defines the requirements of a system that a vendor can provide. It involves reviewing business objectives, developing a vision of how business needs will evolve over time, defining what systems and capabilities a company currently has and how or if they can evolve.

Development of an RFP usually involves interacting with some or all of the departments within a company. Gaining information and commitments from department leaders can be very important to the successful implementation of an RFP. Be sure to plan enough time to not only come up with the RFP document itself, but to brief executives and other stakeholders along the way.

Companies can identify their WiMAX system and service requirements through the RFP development process. These requirements include the types of access devices that can be used, the distribution system that will reach these devices and headend equipment that will gather and reformat content. It also includes middleware software that links access devices to the system, a conditional access system that coordinates and authorizes subscribers access to services, and digital rights management systems that protect content (such as movies) from being copied or misused. Additional requirements include television commerce (t-commerce) applications that allow customers to purchase additional types of products and services along with the customer care, network management and billing systems.

Key RFP Objectives

RFP objectives are statements that identify the desired results that should be achieved as a result of the completion of the project defined by the RFP.

RFP objectives may include determination of the optimal solutions, the selection of a vendor, and the assurance that the selected system(s) and vendor(s) will lead to a successful launch.

Optimal Solutions

Optimal solutions are the selection and implementation of equipment, software and human resources that provide services or benefits that match closest to the objectives of a company or person. Optimal solution criteria may include capital and operational cost per subscriber, the types of services that can be offered and the timeline for installing and deploying the system.

Vendor Selection

Vendor selection is the process of identifying one or more companies to supply products or services. When vendor selection is performed as a result of a request for proposal (RFP), vendor selection may be performed through the use of a criteria evaluation. The criteria evaluation may be itemized and each technical requirement weighted as to the importance of each evaluated item. For large systems, a consultant or system integrator can provide unbiased technical advice during vendor selection.

An eligible vendor is a provider of products or services that has been recognized by the RFP issuing company that meets the qualifications of the RFP process. Vendors usually need to meet qualification requirements to determine if they are likely to be able to perform or provide services or products defined in the RFP. Potential vendors may be asked to complete a prequalification application to determine if they are an eligible vendor, or the issuer may conduct a full RFQ to obtain a list of qualified vendors.

Successful Service Launch

A service launch is the process of making a product available for purchase and distribution with promotional efforts. The criteria that may be used to determine a successful launch can include service quality levels, connection reliability and transmission performance.

Decision Criteria

Decision criteria are a set of evaluation criteria elements (such as price, support and installation time), along with the rating values or processes that are used to assign a value to each criteria element. The decision criteria for company RFP's should be based on the criteria that are likely to help the company reach its business and operational objectives.

Decision criteria can be itemized and weighted as to importance and placed into a decision matrix. A decision matrix is a table (typically in an electronic spreadsheet) or related set of data that contains qualifying rules and quantifying processes (rating) of the criteria that is used to make a decision (such as selecting a vendor to supply a product or service). Business leaders in the organization may work together to define the evaluation process that will be used to determine the winning responder. Having consensus from stakeholders and business leaders can make the implementation of the solution much easier.

The Need for WiMAX

Broadband communication service is the transfer of digital data, including audio (voice) and/or video communications at rates greater than the data rates otherwise available (typically above 1 Mbps per user). Broadband connections allow *multimedia* connectivity to the Internet and applications and provide a rich mixture of data, voice and video for the multitude of applications that increasingly use all three forms of communications. The Internet has created the need for technology such as WiMAX for a vast array of data, voice and video applications and services.

Broadband Private Data Network

Government, commercial enterprises, and small businesses have a need for private data networks. A private data network is typically one that is closed to the outside except for authorized uses. For example, a corporate LAN is only to be used by employees to access company confidential business information, and with only limited access to public data networks (such as the Internet). Wireless LANs offer mobility within buildings and are typically deployed today using the technology developed specifically for short range indoor use, 802.11/WiFi. In the future, WiMAX will be used in conjunction with, or as a replacement of 802.11/WiFi, for WLANs and these WLANs can then be extended to wireless metropolitan area networks (WMANs) across campus and citywide areas. Examples of this include networks for police, military, hospitals, critical infrastructure utilities, and other closely controlled entities. Private networks support their mission with tight security and controlled connectivity.

Broadband Public Networks and Internet Access

Public networks such as the Internet, the PSTN, and Television networks have transformed the way we work, live and play. Consumers and businesses need connectivity to these networks for multimedia applications services and entertainment and want to use these applications in fixed, nomadic, and mobile situations.

These broadband data connections typically allow consumers to connect to public networks at 1 Mbps or above. For businesses using PTP connections, the bandwidth can be engineered to be between 1.5 Mbps (*T1 Speed*) to 45 Mbps (*DS-3)* or higher for links connecting LANs, MANs and WANs to public networks. When WiMAX service providers offer broadband data services that can connect to the Internet, the public network operator may be called a wireless Internet service provider (WISP). If the public network operator offers public telephony access as a service, it may be called a *carrier*, after the regulatory term "*common carrier*".

Digital Telephone and VoIP

Digital telephony is a communication application that digitizes analog audio and transmits the *digitized audio* as a stream of data bits. These analog signals can be audio signals (acoustic sounds) or complex modem signals that represent other forms of information (such as fax signals or *DTMF* / touch tone signals).

WiMAX systems can provide telephone services through the use of IP Telephony (voice over Internet protocol / VoIP). These IP networks initiate, process, and receive voice or digital telephone communications typically using a specialized protocol such as Session Initiation Protocol (SIP)[1]. WiMAX systems can provide digital telephone service through the use of an *analog telephone adapter* (ATA) located on the customer premises, or through the use of IP telephones throughout the premises. ATAs convert IP signals into standard telephone (dial tone) formats.

Digital Television and Video

Digital television is the process or system that transmits video images through the use of digital transmission. The digital transmission is divided into channels for digital video and audio. These digital channels are usually highly compressed. Video compression commonly uses one of the *motion picture experts group* (MPEG) standards to reduce the data transmission rate by a factor of 200:1.

When digital television services are provided by using Internet protocol, it is called IP television (IPTV). IPTV systems initiate, process, and receive television programming using one of the MPEG protocols, over an IP network. WiMAX systems can provide digital television service through the use of *IP set top boxes* (IP STBs). IP STBs convert MPEG over IP signals into standard television formats. Network operators must do careful capacity analysis and planning to ensure that there is sufficient spectrum for the services planned.

The key components of the WiMAX system which include a subscriber station terminal, or subscriber station (SS), a base station (BS) and interconnection gateways to data networks including the Public Switched Telephone Network (PSTN), the Internet, and video content, programming and IP television networks (i.e., IPTV). An antenna and receiver at the home or business SS converts the high frequency (*microwave*) radio signals into broadband data signals for distribution within the business or home.

Figure 3 shows typical services offered by WiMAX providers. The most common is broadband wireless Internet access. Cities are increasingly offering service to citizens and city staff by building municipal networks. While there has recently been a diminished deployment of such networks, due to flawed business models, these networks will continue to be built in the future for the valuable services they provide. Another service available with WiMAX is telephone bypass. In developing countries where telephones are not yet prevalent but *penetration* into the population is growing, governments and entrepreneurs are planning to deploy wireless networks instead of wired networks to deliver telephone services. Even in rural parts of developed countries, wireless broadband service is more cost effective to deploy and operate than wired service. Also, with sufficient spectrum, WiMAX operators can deploy digital television service. With advances in video compression and special design techniques for delivering video to consumers, WiMAX video and wireless IPTV is becoming a reality for some communities. Lastly, with sufficient infrastructure, mobile WiMAX networks can provide broadband wireless to vehicles and public transportation. These networks can compete favorably with PCS and cellular technology, even as they deploy their latest technology for data service.

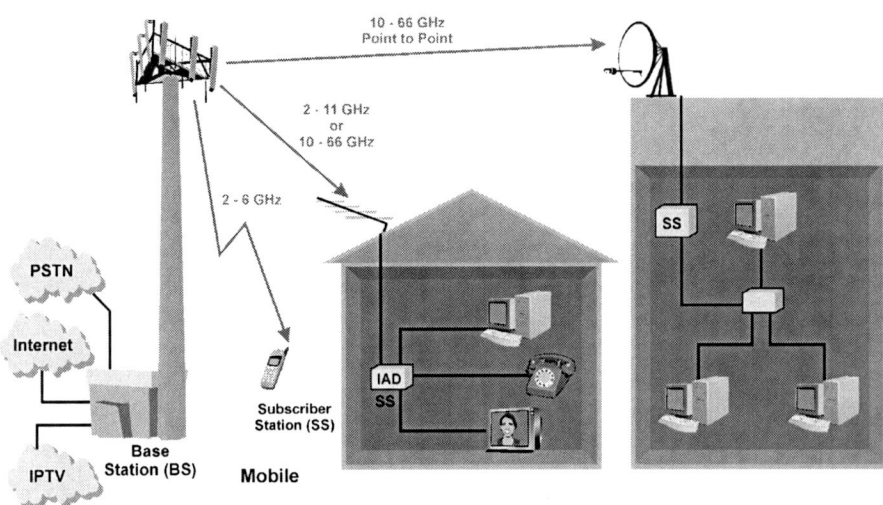

Figure 3: WiMAX Radio System

Figure 4 shows three different *applications* and the network configuration for the WiMAX network. This diagram shows that WiMAX systems can be used for portable (or mobile) services, residential services, or high speed connections for business. For businesses at fixed locations, a WiMAX provider can deliver a service that allows the business to have a high speed "pipe" for Internet access, a private data network, or telephony or video service. A residential service may have a similar set of fixed services, typically competitive with telephone company DSL or cable company data service. Users on foot (or in vehicles) may obtain portable or mobile services at many locations within the service territory.

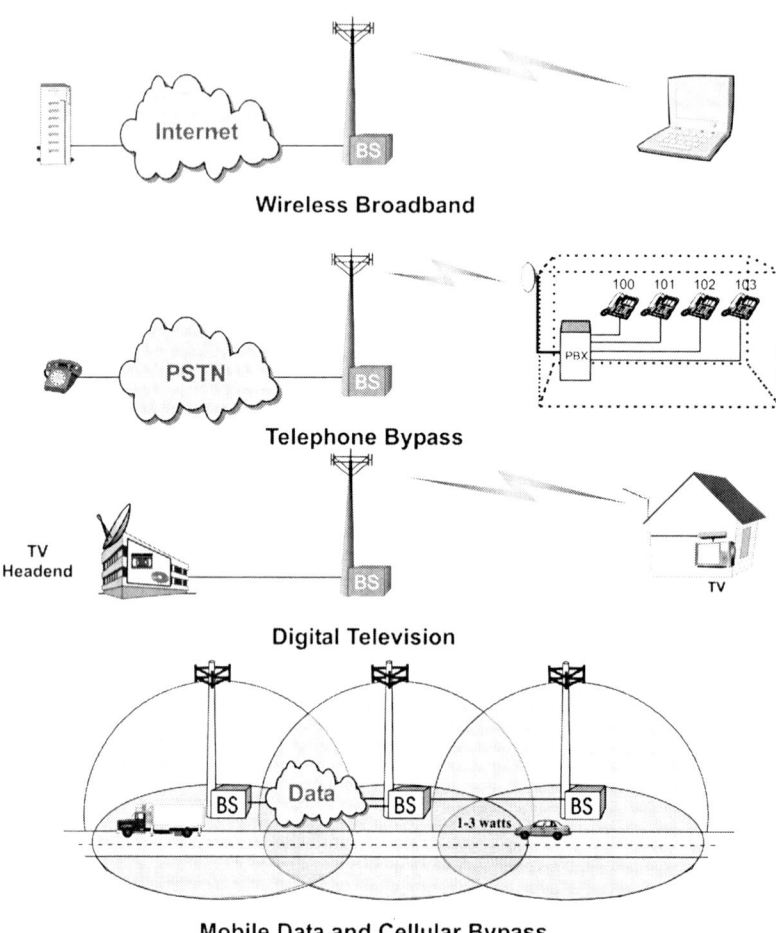

Figure 4: WiMAX Applications

The key benefit of WiMAX over other wireless technologies, including current *cellular data* and *Personal Communication Services* (PCS) systems, is the bandwidth available. The government has set aside the large portions of broadband wireless spectrum, both licensed and unlicensed, that can enable operators to establish a large number of broadband radio channels. A base station that operates using many broadband channels can offer significant bandwidth through the base station. For example, a WiMAX system can provide aggregate data transmission rates that can exceed 120 Mbps [2] under appropriate conditions. This total is then typically divided up per user to allow each of the users to get 1 Mbps bandwidth or more. With the oversubscription typically designed for these networks, this can support 480 to 1200 or more subscribers per base station. This provides two to ten times the bandwidth available for current cellular and PCS technologies. This allows a 'broadband experience' wherever the user may be.

An Overview of WiMAX

WiMAX is a wireless technology certification of a broadband wireless standard. Its purpose is to provide data, voice and video services to public and private enterprises and organizations. WiMAX is based on the fundamentals of wireless, standards, certifications and spectrum. A basic understanding of these principles is necessary to become conversant with important factors of WiMAX.

Wireless

Wireless is a broad industry term that is used to describe communications via electromagnetic waves, and therefore without wires. The communication is carried by these waves from a transmitter to a receiver. Since the communication is typically two-way, the device for communications is called transceiver.

There are several types of wireless communications. They include:

- Land mobile radio (two-way radio)
- Paging and Advanced Messaging (two-way paging)
- Satellite
- Point to point microwave radio
- Family Radio Service, Amateur Radio, Citizens Band radio
- Personal Communications Service (PCS) and cellular radio

These forms of communications provide two way voice, low speed data, messaging communications, transmission, hobbies, and telephone services. They provide communications with another party, or access to a network for a simple application or service.

Broadband Wireless

An advanced form of wireless communications is broadband wireless. Broadband wireless is emerging within some of the wireless networks that are in service today, and it is also emerging in entirely new networks. Broadband wireless communications networks are needed when the application or service requires a broadband communications channel. An example of a broadband application is television video, and a broadband service is broadband Internet access.

Technology Standards

Operational and performance specifications are developed by industry standards bodies to ensure that a technology is built and operates in a uniform way. This allows technology built by one manufacturer to operate identically to that of another manufacturer. The use and conformance to industry standards allows for *interoperability* of gear from different manufacturers.

Certifications

Manufacturers build equipment to technology standards that allow users the benefits of the standards, namely interoperability and standardization of operation from one manufacturer to another (for a certain subset of features). A certifying body is a third party entity that tests manufacturers' equipment to make sure it conforms to the standards, and to ensure that a certain set of features works between equipment of different manufacturers.

Spectrum

The electromagnetic spectrum is the term used for the range of frequencies of electromagnetic radiation. The complete spectrum spans very low frequencies to visible light, and beyond including x-rays and cosmic radiation. A subset of this is the radio spectrum. It spans very low frequencies (30 KHz) also known as long waves, to extremely high frequencies also called microwaves or millimeter waves (90 GHz and beyond).

The radio spectrum is very crowded (although some would say it is very inefficiently used). Governments have regulations and operational requirements that control the use of spectrum. Some spectrum is allocated for a particular service (such as cellular or public safety) and some is allocated for general public use (such as citizens band radio). In developing countries, spectrum may be more available than in developed countries.

Radio System

A radio system is comprised of equipment that can create and manage *radio links* between transceivers. The configuration typically is one central transceiver and multiple remote transceivers, which could correspond to the central location of a service provider and users in remote locations. The central transceiver is usually a large device called a *base station*. The remote units can be smaller and are called *terminals, devices, or users*. If the radio system is used to deliver a service to *subscribers*, the terminals are sometimes called a *subscriber unit*.

Radio and Data Communications

Radio communications is the technology that is used between transceivers to establish a radio communications link. Simple voice or basic data (such as paging) can traverse a radio link through the use of a technique called *modulation*.

Data communications is the general term used for communications involving a stream of (usually broadband) data. Through the use of data communications protocols, complex streams of data can be transported over any communication link. When the link is a radio communications link, it can be called a wireless link.

Networks and Information Technology

Networks enable computers to communicate over data communications links to persons, other computers, and services and applications of all kinds. Networks have traditionally been characterized by their types of use. For example, people use *the public switched telephone network* (*PSTN*) to send and receive information in the audio frequency band.

To setup and manage networks, information technology (IT) systems are used. IT is the study, design, development, implementation, support and management of computer-based information systems, particularly software applications and computer hardware. Increasingly, information technology and methodology are required for planning, designing, operating and maintaining a complex wireless network, either public or private.

WiMAX Standards and Certification

Worldwide Interoperability for Microwave Access (WiMAX) is a *broadband* wireless communication certification that ensures computers and devices of all kinds can connect to high-speed data networks (such as the Internet) using radio waves as the transmission medium, with data rates comparable or exceeding those of other broadband technologies. It is based on a *technology standard*, developed by the international standards body, the Institute of Electrical and Electronic Engineers (IEEE).[3]

The *WiMAX Forum* is the industry association created to certify and promote the compatibility and interoperability of broadband wireless products based upon the IEEE 802.16 (and European *ETSI HiperMAN*) standard [4]. Thus, to help ensure WiMAX products perform correctly and are interoperable with each other, the WiMAX Forum conducts tests to ensure compatibility and interoperability between vendors' equipment. The WiMAX Forum is a non-profit organization that certifies products conform to the industry technology specifications, and interoperate with each other. WiMAX™ is a registered trademark of the WiMAX Alliance and the indication that the product is WiMAX Certified™ indicates products have been tested to the IEEE standard and should be interoperable with other WiMAX certified products, regardless of who manufactured the product.

The WiMAX technology is defined by a group of IEEE 802.16[5] industry standard specifications, and its various revised specifications shown in figure 5, which address particular forms of WiMAX, such as fixed and mobile broadband wireless operation. The fundamental technology used in the 802.16 system evolved from earlier IEEE technology specifications, namely IEEE 802.11 (*wireless LAN*), and even more fundamentally, "Ethernet" IEEE 802.3 (Ethernet).

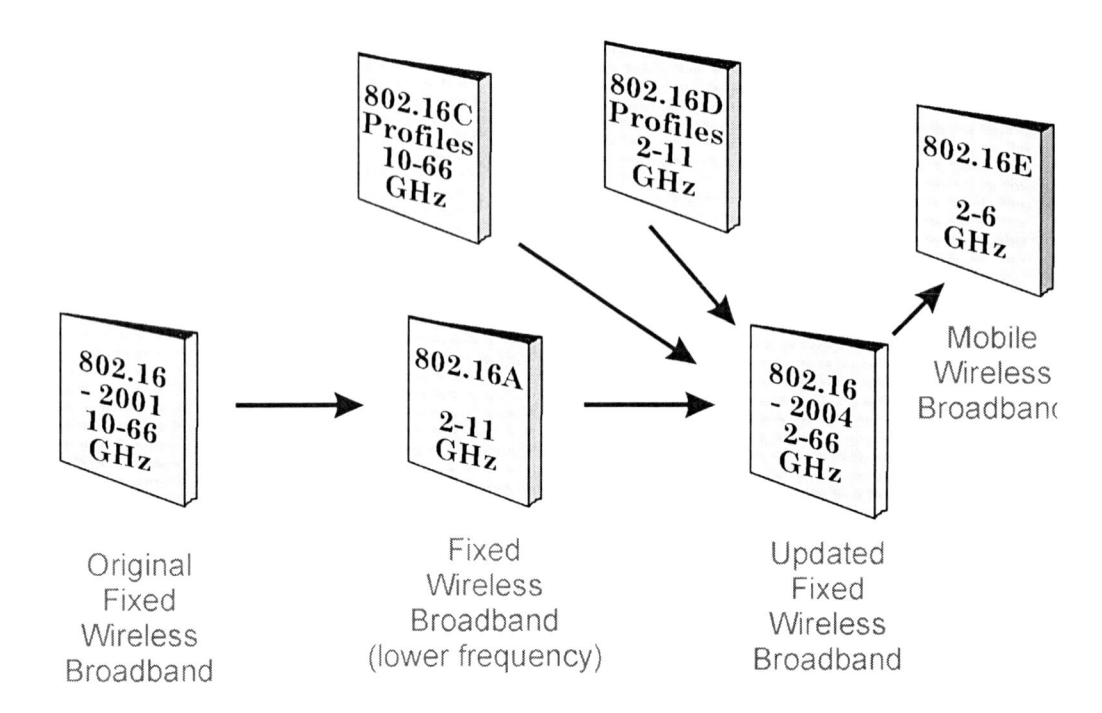

Figure 5: The IEEE 802.16 Standard

At a high level, the term "WiMAX" is frequently used to cover both the IEEE 802.16 technology specifications, as well as the WiMAX Forum's interoperability certification. In some cases, emerging broadband wireless equipment that is designed to the specifications, and is intended to eventually be certified by the WiMAX forum is called "pre-WiMAX".

WiMAX Network

WiMAX is a system that is primarily used as a *wireless metropolitan area network* (WMAN). WMANs can provide broadband data communication access throughout an urban city or geographic area (as well as across large suburban and rural areas). WMANs are used throughout the world and their applications include consumer broadband wireless Internet services, *data transmission lines* (sometimes called *interconnecting lines* or *leased lines*), *Voice over Internet Protocol networks* (VoIP), and transmission of digital television services called IPTV. WiMAX broadband wireless can effectively compete against the wired broadband services provided by telephone and cable companies, such as *DSL service*, and *cable modem service.*

A point to multipoint (P2MP) configuration allows for communications between one central point and multiple homes or offices. The antenna in this case is either an omni-directional or sector antenna which allows some directivity across a wide azimuth and allows the coverage of a large area in one direction.

As shown in figure 6, WiMAX link types have two common layouts or *topologies*; point to point(PTP) and point to multipoint (PTMP). The configuration is useful for carrying a large amount of traffic between two points. The PTP link may be engineered to span a great distance through the use of high gain directional dish antennas. A WiMAX operator could deploy such a *point to point* (PTP) connection or PTP *link* from a base station to a fixed location, using highly directional antennas, which concentrate the signal in one direction, and is typically used to link two locations together. On the left in the figure we see a group of homes connected to an *aggregation or concentration point*, and a PTP transmission link carrying traffic back to the data switch. On the right side of the figure, we see a *point to multipoint* (PTMP) system that allows a tall tower to broadcast the signal, not in a narrow beam to another location, but in a wide beam, to many locations, such as customer homes or offices. PTMP towers are typically 20 to 150 feet tall and provide up to 360 degree *omni-directional* coverage. In order to increase range, add capacity, or control the RF signal, these towers can be sectorized, commonly with 90 degree, 120 degree, or 180 degree sectors. All subscriber antennas are pointed back toward the tower. WiMAX allows security on each individual link that keeps the customers' data separate, private and secure.

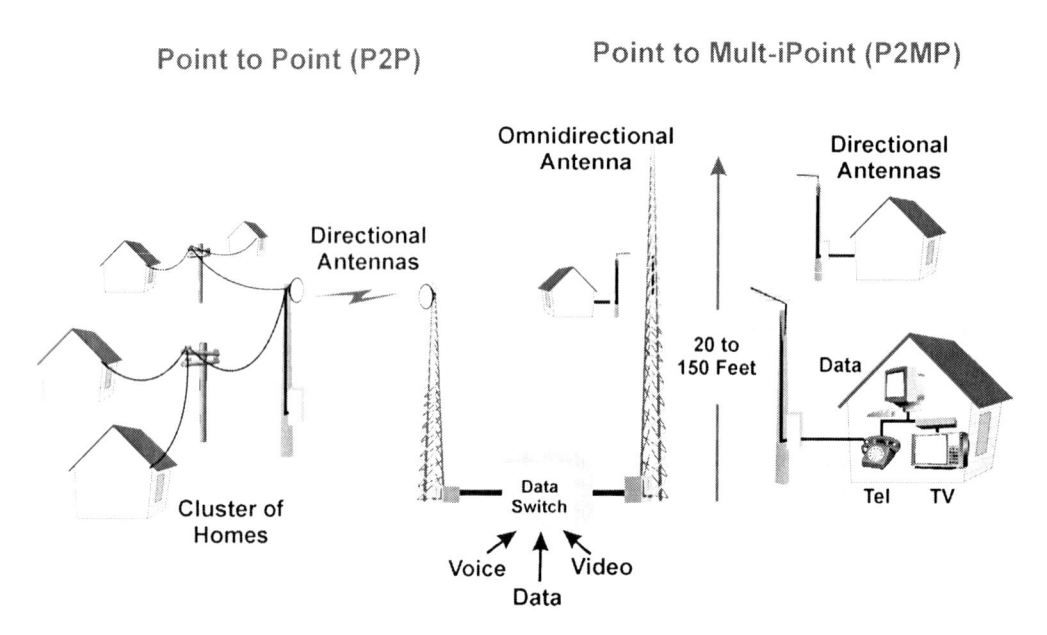

Figure 6: WiMAX Link Types

Fixed Service

The early 802.16 technology was designed for users in one or more fixed locations. To develop a cost effective, high-speed data transmission WMAN system for fixed operation, the IEEE created the first version of the 802.16 industry specification. This early 802.16 system was a short range, *line of sight* system that operated in a very *high frequency* part of the *radio spectrum* (in the 10 GHz to 66 GHz range). This early version of 802.16 allowed service to users by connection of fixed antennas to the outside of their homes or offices.

To allow the 802.16 systems to operate in a lower, longer range portion of the spectrum, the 2 GHz to 11 GHz bands, the 802.16a specification was created.

Nomadic and Portable Service

Nomadic and portable services allow users to move their terminals from one location to another. These "nomadic" users can receive service in multiple locations. While *nomadic service* may be provided in many locations, nomadic service typically requires a transportable communication device (such as a wireless modem or a laptop with a wireless network interface card) to be in a fixed location during the usage of communication service (for example, the user cannot be moving in a vehicle). Consumers are becoming familiar with this mode of operation by using short range *WiFi* connected devices at public and private *hotspots*. WiMAX will have much larger "spots" of coverage, but like WiFi, nomadic service will require users to be stationary during usage. Fixed service has important applications for many types of network operations.

Mobile Service

For mobile service, the *IEEE 802.16e* was developed. The 802.16e specification adds the ability to move at vehicular speeds by the addition of *handoff* (call transfer) and *mobility management*. The 802.16e standard also advanced security features by adding *extensible authentication protocol (EAP)*, and features for mobility such as *power saving sleep modes* to previous versions of the standard.

WiMAX Spectrum

WiMAX has several different physical radio transmission options which allow WiMAX systems to be deployed in areas with different regulatory requirements and frequency availability. The WiMAX system was designed with the ability to be used in *licensed* or *unlicensed* frequency bands using narrow or wide *frequency channels*.

The issuer of a WiMAX RFP has likely made a decision regarding the spectrum in which the WiMAX network will operate. Many operators, such as public common carriers or public safety agencies, have already been allocated or won spectrum rights at a government spectrum auction or through a licensing process and the decision of what spectrum to operate in, is fete accompli. If so, members of the RFP team need not focus on spectrum considerations other than the type and amount description to be listed in the RFP.

Spectrum Due-Diligence

If a spectrum operating band is not pre determined and if there is some flexibility on what band to build a WiMAX network on, then significant analysis, legal and regulatory due-diligence, and performance studies may need to be undertaken to determine the optimum choice of operating band. The decision on operating band is critical in terms of cost, schedule, the types of services that can be offered (particularly video and mobility) and the resulting quality of performance of the resulting network, both at the start of operations and throughout the network's life cycle.

What spectrum and whether to operate in a licensed or unlicensed band is a fundamental, profound decision, as is the selection of real estate for the operation of a traditional business. The decision about operating bands is so critical and so particular to a geographic area, that savvy vendors actually screen out naive customers on first contact by asking "Do you have spectrum?" And if the answer is "no" or "not yet" then vendors of WiMAX equipment that operate in licensed spectrum will not likely spend much time with the potential customer until this aspect (and another fundamental condition like funding) is firm.

License Exempt Spectrum

What if there is no way of securing licensed spectrum, either due to the cost or availability? The answer is to use unlicensed, or more precisely, license exempt spectrum. Various regulatory bodies have set aside spectrum for use by operators in a shared fashion, without the need for (nor given the protection and exclusive use of) licensed spectrum. Shared, license exempt spectrum is becoming more and more available around the world due to the strategic trends of broadband multimedia data needs. If the strategy to deploy WiMAX is to use unlicensed spectrum, the operator will need to do engineering studies of the spectrum in the areas of proposed operation as soon as possible, along with a study of likely future sharing of the spectrum and its impact on bandwidth, performance and quality of the services and applications offered.

Regulations vary based on the specific part of the world, country, and even – just like real estate — the region or zone within a region, state or city (for example in a new 3.6 GHz band in the US, certain parts of the country are grandfathered for military radar systems and new operations such as WiMAX must not interfere in those bands, which has an impact on the equipment selection and services to be offered in the area).

Spectrum Details for a particular region

For both licensed and unlicensed spectrum, the details of exact spectrum bands, regulations regarding operating modes such as fixed, portable and mobile, and the details of implementation in specific regions and for specific scenarios must all be determined, at least at a high level, prior to setting down technical requirements in a WiMAX RFP. Frequently, an operator will seek information from the vendor community on feasible and practical network strategies regarding these regulatory issues by issuing an RFI to gather high level information suitable for finalizing a strategy. This can be done before issuing a full blown RFP that seeks a final quotation for a very specific network design and architecture that uses a particular spectrum allocation.

Spectrum Available for WiMAX

The spectrum available for WiMAX varies widely by country. Also, regulatory bodies in each country are at various stages of regulating spectrum for WiMAX (as well as other wireless technologies including PCS). Currently the most available licensed bands are 2.3 GHz, 2.5 GHz, and 3.5 GHz. In the unlicensed bands, the 5 GHz band is generally available worldwide. Figure 7 shows the WiMAX profile information for several popular WiMAX bands.

Band Information	2.300-2.400	2.500-2.690	3.400-3.600	5.725-5.850
802.16-2004 WiMAX Profile	No	No	Yes	Yes (TDD only)
802.16e-2005 WiMAX Profile	Yes (TDD only)	Yes (TDD only)	Yes (TDD only)	No
Channelization	8.75 MHz	3 MHz, 5.5 MHz	3.5 MHz, 7 MHz	10 MHz
Licensed or License Exempt	Licensed	Licensed	Licensed	License Exempt
Spectrum allocation per operator	27 MHz	3x5.5 MHz + 6 MHz (total 22.5 MHz)	From 2x7 MHz to 2x56 MHz	Not Applicable
Where available	South Korea	US, Mexico, Brazil, Southeast Asia, UK, Canada, Australia, Central America	Worldwide except US	North America, South America

Figure 7: Information per frequency band

Other bands are currently under consideration for use as future broadband wireless bands potentially suitable for WiMAX network deployment. The list of potential bands includes:

The 700 MHz Band

Broadcast Television License-exempt spectrum

902-928 MHz

2.40 - 2.4835 GHz

24 GHz band

Millimeter wave bands at 60, 70, 80 and 90 GHz

Developments must be closely watched within each operating region to determine strategically whether or not a particular band is emerging appropriately for WiMAX in terms of the right regulatory (fixed or mobile, power limits, etc.) and technological (i.e. vendors interested in building products for the band) conditions for successful WiMAX operations.

WiMAX Certification

For certified interoperability of devices and for a basic level of multivendor operation, the use of WiMAX certified equipment is required. In order to certify that equipment is interoperable and in compliance with the IEEE 802.16 technology standard, WiMAX certification laboratories around the world test one manufacturer's device(s) against other manufacturer's device(s) to ensure that a basic subset of features and configuration defined in the technology standard are implemented and operate properly. Figure 8 shows that the WiMAX forum has focused on three main bands, 5.8GHz,

System profiles	Certification profiles		
	Spectrum	Duplexing	Channel width
Fixed WiMAX (IEEE 802.16-2004, OFDM)	3.5 GHz	TDD	3.5 MHz
	3.5 GHz	TDD	7 MHz
	3.5 GHz	FDD	3.5 MHz
	3.5 GHz	FDD	7 MHz
	5.8 GHz	TDD	10 MHz
Mobile WiMAX (*) (IEEE 802.16-2005, OFDMA)	2.3-2.4 GHz	TDD	3.75, 5, 10 MHz
	3.4-3.8 GHz	TDD	5, 7, 10 MHz
	2.496-2.69 GHz	TDD	5, 10 MHz
	3.3-3.4 GHz	TDD	5, 7 MHz
	2.305-2.320, 2.345-2.360 GHz	TDD	3.5, 5, 10 MHz

Figure 8: WiMAX Profiles - System and Certification
Source: "WiMAX and IMT-2000," Reprinted with permission of the WiMAX Forum

3.5GHz and 2.5GHz for conformance testing and interoperability testing as a start (even though the IEEE 802.16 technology standard is targeted for use at any frequency below 11 GHz that can be channelized between 1.75 MHz and 10 MHz).

WiMAX Profiles

The WiMAX Forum has defined both system profiles (Fixed, Mobile, etc.) and certification profiles (band, channel size, duplex method). System profiles address fundamental operating modes while the certification profiles group features and configurations into a testable structure. In doing this, they selected popular configurations with suitable spectrum around the world in operating bands (e.g., 3.5 GHz, 5.8 GHz, etc.), channel bandwidths (e.g., 3.5 MHz, 5 MHz, 7 MHz, etc.), and duplexing protocols (TDD or FDD, time or frequency division duplex). The units under test must conform to the 802.16 standard, as well as operate among themselves successfully. Thus, a WiMAX certified piece of equipment conforms and is (only) interoperable with equipment under a particular system and certification profile. A large group of new tests are added in each new test case Release to WiMAX certification testing, and as new Releases are developed, incrementally to earlier Releases, vendors must re-test their product to ensure their product successfully interoperates and conforms. Doing so, they can continue to enjoy WiMAX Certification (prior tests do not change thus preserving backward compatibility). Waves are a smaller group of new tests that come between major Releases. For example, the Fixed WiMAX system profile has a Release 1.0 and Wave 1 (covering mandatory, fundamental features of the standard) and Wave 2 (covering optional features such as Quality of Service and AES Encryption). Figure 9 shows the WiMAX certification release process.

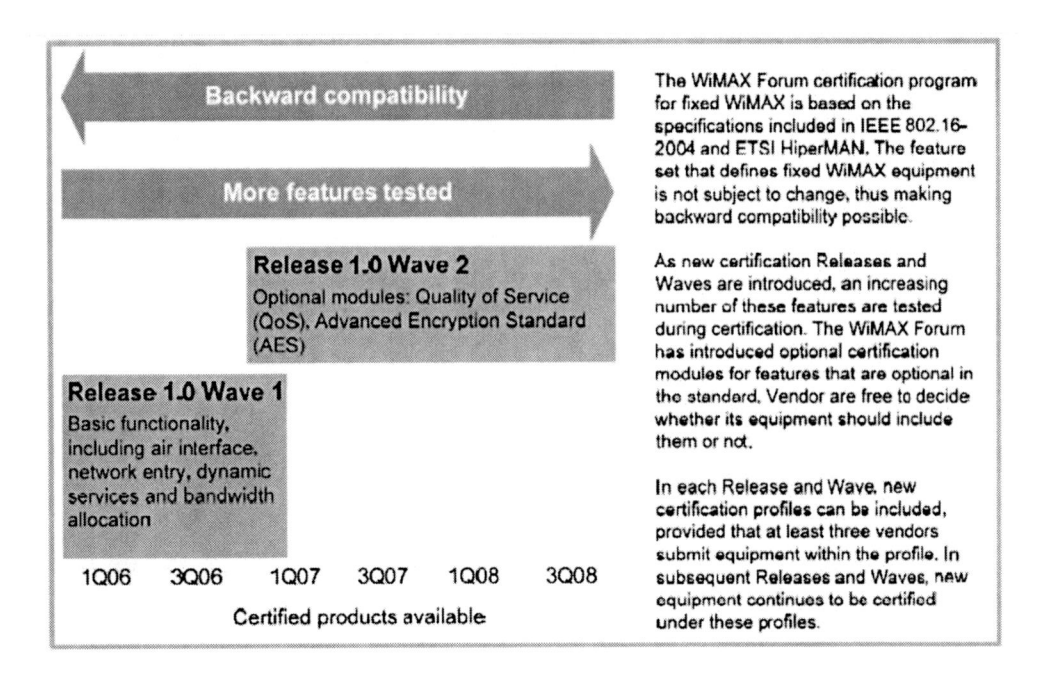

Figure 9: WiMAX Certification Release Process
Source: "The WiMAX Forum Certified™ Program for Fixed WiMAX™," Reprinted with permission of the WiMAX Forum

Profiles Related to WiFi

Figure 10 depicts the current and proposed, licensed and unlicensed, WiMAX profiles as well as how they relate to WiFi frequencies.

Licensed		Unlicensed	
2305-2320 MHz	US WCS	2400-2480 MHz	ISM (802.11 Wi-Fi)
2345-2360 MHz	US WCS	3300-3400 MHz	(future)
3400-3600 MHz	3.5 GHz band	5150-5350 MHz	U-NII Band (Wi-Fi)
		5470-5725 MHz	WRC
		5725-5850 MHz	U-NII/ISM

Figure 10: WiMAX Profile Frequency Bands

Certification Not Wanted

While the operating band decision is critical with regard to spectrum to be made by an operator of a WiMAX network, perhaps an even higher level decision is if an operator even needs WiMAX certification, which allows multivendor competition for infrastructure purchases, and standardized operation for a wide variety of user devices. For example a public carrier deploying a city-wide WiMAX network would clearly want any potential consumer with a WiMAX certified device (say a laptop from any retail outlet) to be able to attach to the network. On the other hand, a private enterprise, public safety organization, public utility, hospital or military agency may not need or even want multi-vendor infrastructure or non-agency devices attaching to the network thus making WiMAX certification less important.

WiMAX Requirements

When an operator or carrier of a private or public network is planning a WiMAX system deployment, an RFP is typically an important step in the process. An RFP is filled with the technical requirements that the operator or carrier has for the network.

Requirements are divided up into business and technical requirements. Business requirements include financial and marketing requirements. Technical requirements include performance and quality requirements. Collectively, these requirements enable the issuing operator or carrier to specify what is needed, so that RFP respondents provide exactly that.

WiMAX Business Requirements

Applications, Services and Use Cases Requirements

WiMAX is a broadband wireless technology that can provide a broadband service or data pipe to users. Some applications for WiMAX as a data pipe include:

- Wireless Broadband Internet access
- DSL or cable modem fill-in, reach-out, or replacement
- Wireless backhaul for PCS and WiFi Hotspot providers
- T1 or ISDN circuit replacement for business
- Second data line for small offices or home offices
- Mobile broadband data services
- Private mobile network
- Rural broadband coverage
- Security, surveillance, telematics, telemetry, SCADA

These data services are growing and WiMAX can provide "big pipe" connectivity to numerous small businesses, enterprises and municipal organizations.

Voice service is another application that WiMAX can provide. Organizations such as clinics, schools, government buildings, and social centers all have a need for the replacement of expensive leased lines. Emerging virtual PBX (VPBX) services can now offer all of the telephone features only previously available through telephone company Centrex services or through the purchase of a physical PBX.

WiMAX can deliver video, given sufficient spectrum, to remote locations and provide a video service to consumers and businesses. Video is a bandwidth intensive application, so WiMAX may not be able to deliver the number of simultaneous video channels available through a cable provider. Careful network capacity planning is necessary. However, WiMAX may be able to deliver new video services that customers will find attractive and useful such as video conferencing, mobile video, and streaming multimedia.

Use -Cases

In order to create appropriate business requirements that allow RFP respondents to design an effective network, RFP issuers usually develop use cases for the most important applications or services to be offered. A use-case is a clear description of an application or service from a user's perspective. The use case describes in detail the steps the user takes from start through delivery of the service to completion. Use cases uncover business requirements that may not otherwise be articulated.

Service Area & Coverage Requirements

The service area of a WiMAX network is the area where WiMAX service is available from the network operator's base stations. This is a very important business requirement and is determined by the network operator after much analysis regarding competition, demographics, user mobility, financial and cost considerations, and many other details.

The coverage requirement for business purposes can be campus, city, regional, or national. It can be outdoor and/or indoor. It can be fixed, portable and/or mobile. It can be urban, suburban, rural or remote.

Marketing and Selling Requirements

If the network to be built is in competition with other providers, the competitive business environment requires that the marketing function carefully analyze the existing services in the market, and provides competitive service capabilities, bundles, pricing, and coverage in order to establish a viable business.

Finance and Accounting Requirements

The finance and accounting of a WiMAX network must allow for price competitiveness, cost control and predictable operations and maintenance costs throughout the life of the network in order for it to be viable.

Careful strategic *techno-economic analysis* must be conducted to ensure the network remains financially viable under numerous predictable and unpredictable circumstances. *Financial modeling* using a financial business case developed in spreadsheet format allows for "what-if" analysis of potential "future states" or various *scenarios,* to understand the dynamics of a network's financial performance. Typically an RFP response from vendors provides important detailed cost information that is useful for the financial planning of a WiMAX solution.

Also important is the tactical accounting information needed from the WiMAX network solution. This would aid the accounting function efficiently and effectively perform internal and external billing and accounting activities.

Business to Technical Requirements

Figure 11 shows the typical requirements of a WiMAX system along with the translation to a class of technical requirements that must be described by the RFP issuer. Also shown are the classes of technical specifications that the RFP respondents must address in a solution. Organizing and describing these and similar requirements are the subject of this book. The solution is provided by each individual RFP respondent, in a detailed design and architecture, listing technical specifications of the solution proposed.

Business Requirement	Technical Requirement	Solution
High Data Rate and Bandwidth	High spectral Efficiency and low self and neighbor interference	Cellular design, adaptive modulation and coding, OFDM modulation, Adaptive Antennas, Sectorization, Dynamic channel assignment
Indoor coverage	Non-line of sight operation	Effective use of multipath through diversity, OFDM modulation, spread spectrum, etc.
Low Cost	Standards based network technology	Use of Internet Protocol (IP) standards as well as IEEE (802.16) and WiMAX certification for purchase of network elements and subscriber units produced in high volume
Mobility	Contiguous coverage, good handoff performance.	Session continuity, roaming continuity, seamless handover, mobility management databases, location updating, mobile IP techniques
Portability	Long battery life, small size, light weight, non line of sight operation	Modulation that allows for sleep and idle modes, low power circuitry
Priority for certain users	Quality of Service	Coding that supports voice, video and data. Priority for classes of users and classes of applications per user. Radio resource management, efficient scheduling schemes, IP level QoS
Security	Authorization, Privacy, encryption, and data integrity	Authentication, Authorization and Accounting functions, strong encryption, integrity of user transmissions and data.

Figure 11: Business to Technical Requirements

WiMAX Technical Requirements

Technical requirements take higher level business requirements, which can be more qualitative in nature, and provide more distinctions suitable for building and testing the system by detailing performance or quality attributes and assigning quantitative information to these attributes. For example, a business requirement may be that the WiMAX system operates to units indoors as well as units that are mobile. The technical requirements would then specify, for example the percentage of the locations and the percentage of time the service would operate successfully inside a typical structure such as a frame home or office building. And it would specify the minimum speed at which the service would continue to operate for a subscriber unit located in a typical vehicle.

Service and Coverage Requirements

The WiMAX service is usually defined in terms of applications within certain wireless service *coverage*. At the business requirement level for a particular WiMAX service, a requirement could be stated, "The service shall allow for Internet access within the geographic boundaries of the city of X." This is a good requirement for informing the respondent of the type of service (such as voice, data, or video) and where it is to be offered. The technical requirements would elaborate on, and deepen the above general requirement for coverage. The general requirement is typically known as, the *coverage objective*. Figure 12 shows an example coverage objective for a WiMAX system.

Figure 12: Wireless Coverage Objective
Source: EDX, www.EDX.com

Coverage objectives (requirements) provide the respondent with a target within which to design and quote a WiMAX system solution. The issuer must address three types or levels of coverage: 1) the *extent of coverage*, 2) *continuity of coverage*, and 3) the *depth of coverage*. These three levels or types of coverage must be defined in order to properly design the solution. Figure 13 shows how the extent and continuity of coverage area for a WiMAX system might vary from its design objectives.

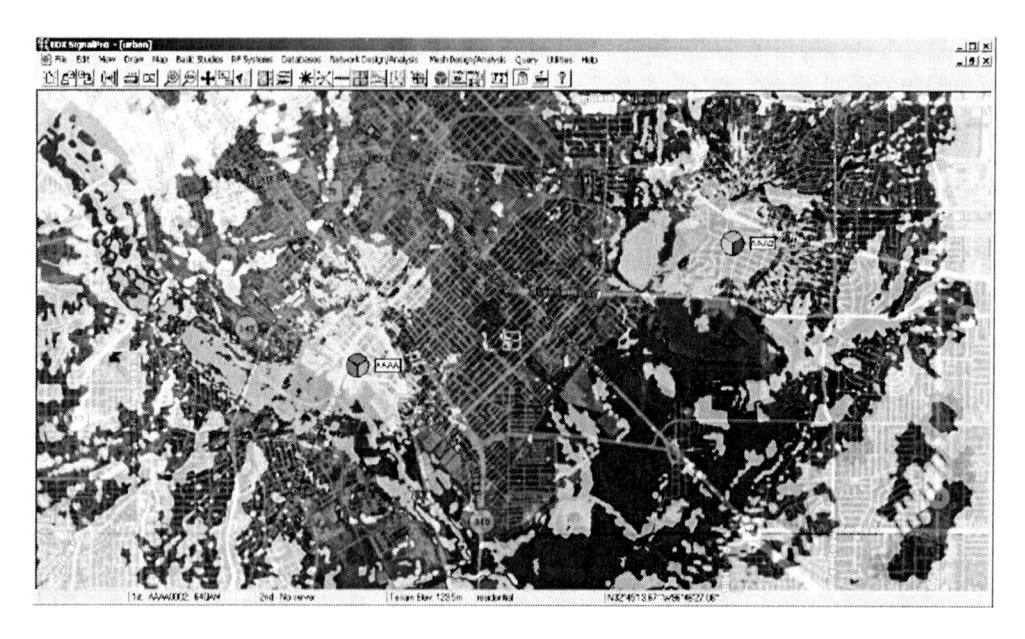

Figure 13: Real or Actual Wireless Coverage
Source: EDX, www.EDX.com

Extent of Coverage

Extent of coverage refers to the expanse of geography to be covered. Continuity of coverage refers to the uniformity of coverage across a region, whether contiguous or discontiguous, which is important for mobile operation versus fixed operation. Depth of coverage is the signal strength level that is needed for penetration into the outdoor tree or clutter *canopy*, or indoor building environments, or inside vehicle units.

Wireless coverage is characterized, predicted and measured using statistical techniques. This is because of the huge number of variables that go into an accurate description, prediction or realization of a real radio signal in terms of the quality of the radio signal at a specific place and time, or during movement in and around *obstacles* and *clutter* in the environment. There is no way to avoid the use of statistical methods when talking about or establishing a *statement of work* (SOW) with *contractual* requirements for, wireless coverage.

Range of Coverage

The primary factor that affects coverage, or *range*, is the environment. Hills, buildings and trees, or in general *clutter* or *obstacles*, affect *radio signal propagation*, and can mean the difference in range between a base station to user station of less than one mile to well over 30 miles. This is why one hears such diverse claims for wireless (and WiMAX) system coverage, if one neglects to mention the intervening clutter, or depth of coverage, one can assert the range of a station to be very long (probably with high gain fixed antennas with LOS paths) or very short (likely assuming indoor coverage, through lots of clutter, with simple consumer antennas, and devices that "should be able to operate anywhere I want").

Bandwidth and Throughput

Bandwidth and throughput define the number of bits, characters, or blocks of data that a system working at maximum speed can process during a specified period of time. The throughput varies with the type of modulation that is used.

Figure 14 depicts the various throughputs in Mbps that are achievable with WiMAX using the *modulation types* shown. Modulation is the process for encoding information onto the transmitted signal. The more bits of information that can be encoded onto a symbol of transmitted information, the more concentrated the information rate of the signal, and the more *throughput* provided by the radio link. The types of modulation range from very concentrated 256 levels *Quadrature Amplitude Modulation* (QAM) to the least concentrated *Bi-Polar Shift Keying* (BPSK). As the signal becomes weak at long distances from the transmitter, the less concentrated modulation type must be used. While the throughput goes down, this allows WiMAX to communicate even at long distances.

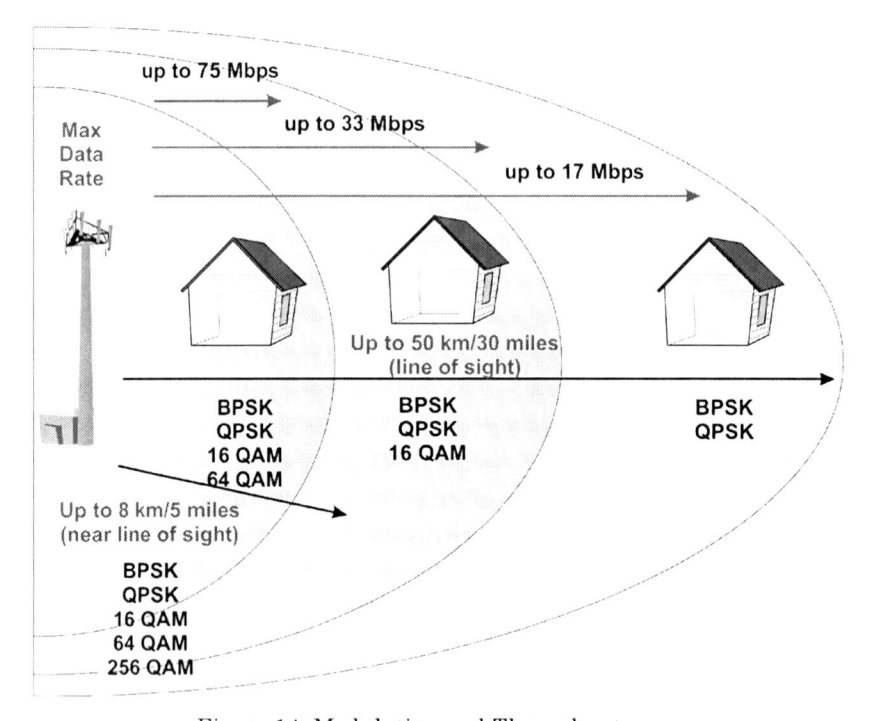

Figure 14: Modulation and Throughput

Types of Coverage Levels and Location Availability

The type of requirement and level specified depends on the applications to be delivered. Also, many times, there is a geometric relationship between "goodness" and" total cost ", particularly with wireless coverage. For example, the cost of coverage quality might double to cover from 50% to 90% of the locations within a region, or *location availability*, and it might double again to increase the requirement from 90% to 95%, and it might once again double from 95% to 98%, with continued diminishing returns in coverage quality as you attempt to achieve 100% coverage just by adding infrastructure/cost. Therefore, the exact type and level of coverage is difficult to provide in an absolute way, rather it depends heavily on the specific circumstances, applications, and environment – the more the better, until you reach diminishing returns.

Cost of Coverage

It is often of interest to understand the cost of coverage. Issuers may request a series of coverage levels; say 90%, 95%, and 99% in an RFP in order to understand the cost of coverage. By requiring respondents to provide coverage designs at a few levels, issuers may be able to judge when coverage availability is "good enough" (for the cost).

Outdoor Coverage

The extent of outdoor coverage requirements can be specified as a functional or an absolute requirement. A functional requirement might be, for example, cover the City of Dallas out to the city (geographic) limits.

Indoor Coverage

The extent of indoor coverage is typically specified as a percentage of the locations as well as a percentage of the time. In addition to that, the actual practical requirement varies greatly depending on exactly how the operator plans to cover the interior of a building. If the operator has access to the building and can place base stations, access points, *pico-cells*, or *femto-cells* within the building, the coverage requirement might be on the high side and uniform throughout. For example the requirement might state, "Coverage must include 80% of the habitable locations within the building". If the operator does not have access to emplace equipment inside the building, but must *illuminate* the building from the outside, then for practical purposes, the specification is typically relaxed and non-uniform; otherwise one quickly approaches diminishing returns. The requirement might state, "The respondent shall design for indoor coverage to 50% of the locations, 50% of the time to all interior space containing one exterior wall and containing one exterior facing window." In this case, it might not be prudent to require signal strength to interior offices where the signal would have to penetrate two or more walls, or a wall without a window, since the cost would likely be prohibitive to cover such space, through the use of only exterior base stations. For this particular case, it might be necessary to obtain an agreement with the building owner before an RFP is issued that would allow the operator to place a base station or a *distributed antenna system* (DAS) inside the building, and inform the RFP recipients that this approach is available.

Let us note that municipal network operators have the dilemma of not generally being able to gain access to interior spaces of privately owned homes and offices. Cities and utilities typically have lots of places where they can mount APs or base stations, such as telephone company poles, city light poles, electric substations, or other *mounting assets,* etc. Private network operators have the opposite dilemma; they can place APs inside and on their property, but they do not have mounting assets across the city where they can extend coverage outside and away from their property. Some consortiums have formed that bring together licensed spectrum, mounting assets, and a contract for service with the ability to place a DAS or APs within private property in exchange for a contract for preferred service and gain good indoor as well as regional outdoor service.

Link Budget

Link budget is the maximum amount of signal losses that may occur between a transmitter and receiver to achieve an adequate signal quality level. The link budget typically includes cable losses, antenna conversion efficiency, propagation path loss, and fade margin.

Respondents should provide calculations that support their coverage design. The basis for the coverage design is the range of the radio link. The radio link range is determined by the *system gain*, or the *link budget* of the equipment. Figure 15 shows the typical parameters for a link budget of a WiMAX system. The values of all parameters must be specified by the vendor as specifications of the coverage design for a WiMAX system.

Link Budget Parameter		Units	
Transmitter output power		dBmW	
Circulator / hybrid loss		dB	
Coax or waveguide loss		dB	
Antenna radome loss		dB	
Antenna gain		dBi	
Effective Radiated Power		dBmW	
Receiver antenna gain		dB	
Antenna radome loss		dB	
Coax or waveguide loss		dB	
Subchannelization gain		dB	
Antenna gain		dBi	
Circulator / hybrid loss		dB	
Effective Receiver Signal Strength		dBmW	
Receiver noise level		dBm	
Channel bandwidth		MHz	
Receiver noise figure		dB	
Required Signal to Noise ratio, SNR		dB	
Required Signal to Noise plus Interference ratio, SNIR		dB	
Required thermal noise fade margin		dB	
Required interference fade margin		dB	
Total system gain (or link budget)		dB	

Figure 15: Link budget showing system gain

Site Surveys

Respondents may have a need to perform a site survey before, during and/or after responding to an RFP. An initial site visit may, as the primary purpose, perform inspection of the sites available for placement of base stations and network management platforms at the network operations center. A field strength measurement of the noise floor and interference may be taken at this visit or at a subsequent visit during preparation of the RFP response. To gain the most accuracy in the RF prediction of coverage, the respondent or winning vendor will perform detailed measurements that calibrate their prediction tools to achieve the most accurate radio coverage design possible. Since this is usually a costly engineering undertaking, with drive testing and measurements throughout the coverage objective, this level of site survey may take place after a winner is selected in the RFP process and as part of contract negotiations or initial implementation to fully characterize the coverage to be provided.

Data Network Requirements

WiMAX is primarily a data networking technology to connect large amounts of digital data across *"the last mile"* from the *network infrastructure* to where users use the service.

Data network requirements include attributes and values that characterize the performance and quality of the data transmission, such as the maximum level of delay or the maximum number of errors tolerated. It is important to specify enough measurable criteria to ensure proper operation, performance and quality of the application or service envisioned. Furthermore, a measurable requirement allows the operator to test the system before acceptance, and have the vendor prove the system meets the requirements set forth by the operator.

WIMAX system requirements could address both *legacy* type services, such as existing telephone service, as well as *state of the art* new services such as IPTV from traditional network television programming sources, or even emerging integrated television commerce (*t-commerce*) capabilities.

Internet Access Requirements

Internet access requirements define the capabilities for a user or device to connect to the Internet through the WIMAX network. WIMAX systems may offer a combination of services including Internet access. Internet access through a WiMAX system is usually controlled and compartmentalized to ensure that Internet services do not interfere with television services or telephony services.

The RFP responder should describe how they meet the operator's requirements in terms of access link latency, gross and net throughput for Internet traffic for individual users as well as aggregate out of the base station.

Typically, the operator procures the Internet gateway service from a commercial Internet Service Provider in order to connect the WiMAX network to the Internet for customers.

Besides WiMAX infrastructure equipment, there exists Internet Service Provider platforms that must be procured through a separate RFP or system integrator, in order to provide Internet access as a service to subscribers, including business support systems such as billing platforms, subscriber management platforms, and provisioning systems.

Figure 16 shows the many requirements that are commonly considered when creating a request for proposal (RFP) for a WiMAX system. This diagram shows that the requirements include the service coverage area, whether the network is for fixed, portable or mobile service, requirements for operations and maintenance, and service and application requirements including Internet access, VoIP and IPTV video capabilities.

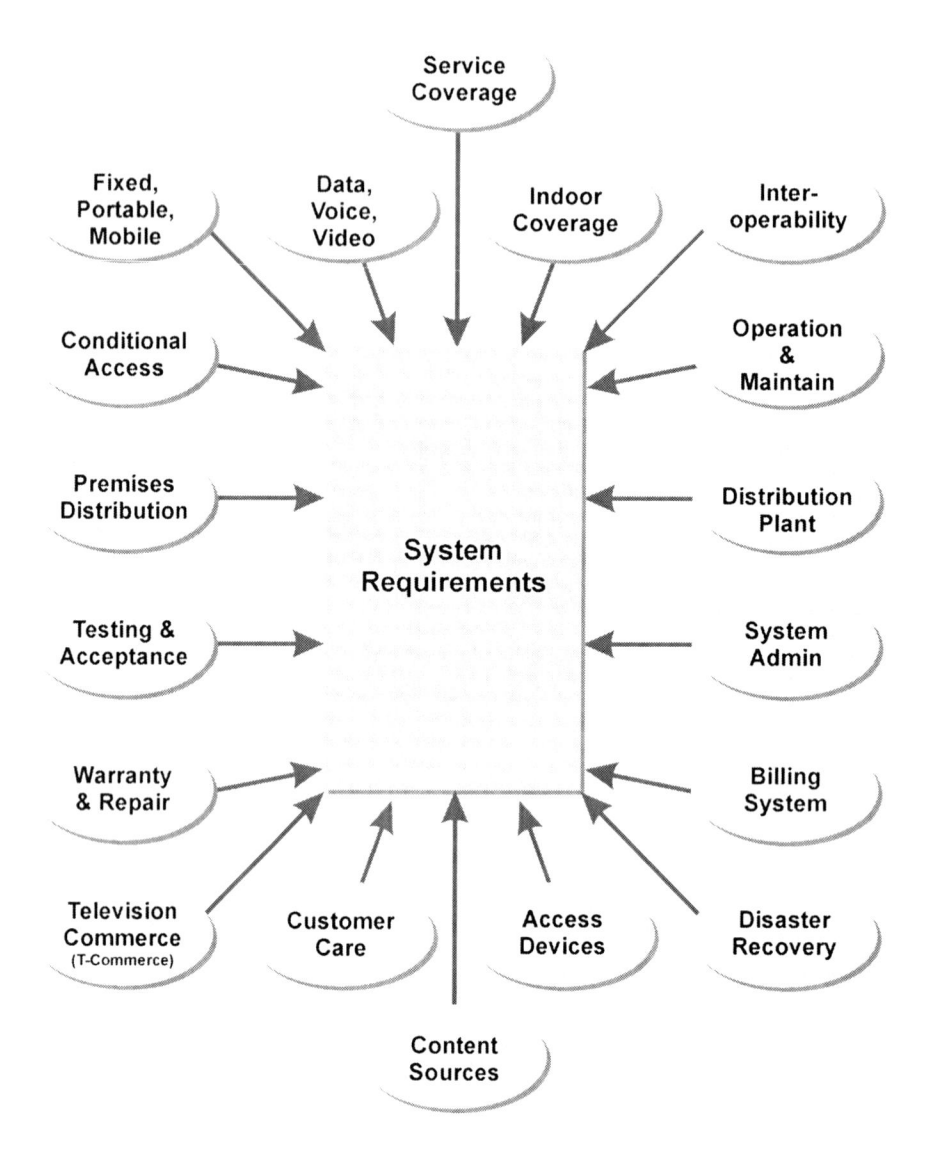

Figure 16: WiMAX RFP Requirements

Voice and Telephony Requirements

Telephony requirements are the capabilities of a user or device to connect to the *public switched telephone network* PSTN through the WIMAX network. WIMAX systems may offer a combination of services including residential type second line service, to fax line, to business class CENTREX type services using modern *Virtual PBX* services. Telephone access through a WiMAX system is usually controlled to ensure that telephony services do not interfere with other services such as television or Internet data access services.

The RFP responder should provide a response to the operator's requirements for telephone service, including number of lines, CLASS type services available and total number of VoIP lines per base station. Typically, the operator procures a VoIP gateway service provider for connection of VoIP customers to the PSTN.

In addition to WiMAX infrastructure equipment, VoIP service platforms or a Virtual PBX service must be procured through a separate RFP or system integrator, in order to provide CLASS services or CENTREX type services to customers. Also, typical ISP services are needed including billing platforms, subscriber management platforms, and provisioning systems.

Video and Television Requirements

Television services include video programming and content to multimedia viewing devices (e.g. televisions, media center computers, etc.). The television service requires the delivery of television video signals over a digital transmission medium that can be received, decoded and rendered on a viewing device in the home or office.

Typically, the operator procures for the programming content services separately from a programming content service provider in order to secure quality programming for customers.

In addition to WiMAX infrastructure equipment, Video processing and storage platforms for management of video content and programming for customers must be procured through a separate RFP or system integrator due to the specialized nature of these capabilities. They include not only billing platforms, subscriber management platforms, and provisioning systems, but also encoding platforms (MPEG-4, MPEG-2 or VC-1), media storage and switching devices, and video format types (profiles such as standard definition (SD) and high definition (HD)).

Access Devices Requirements

Access devices are any type of equipment that can be connected at the end of a communications system or circuit to access the network. Access device types include telephone sets, television sets and computers that have network access. Figure 17 shows the types of WiMAX access devices that may be used in a WiMAX system, including external radio modules, PC cards and WiMAX embedded devices.

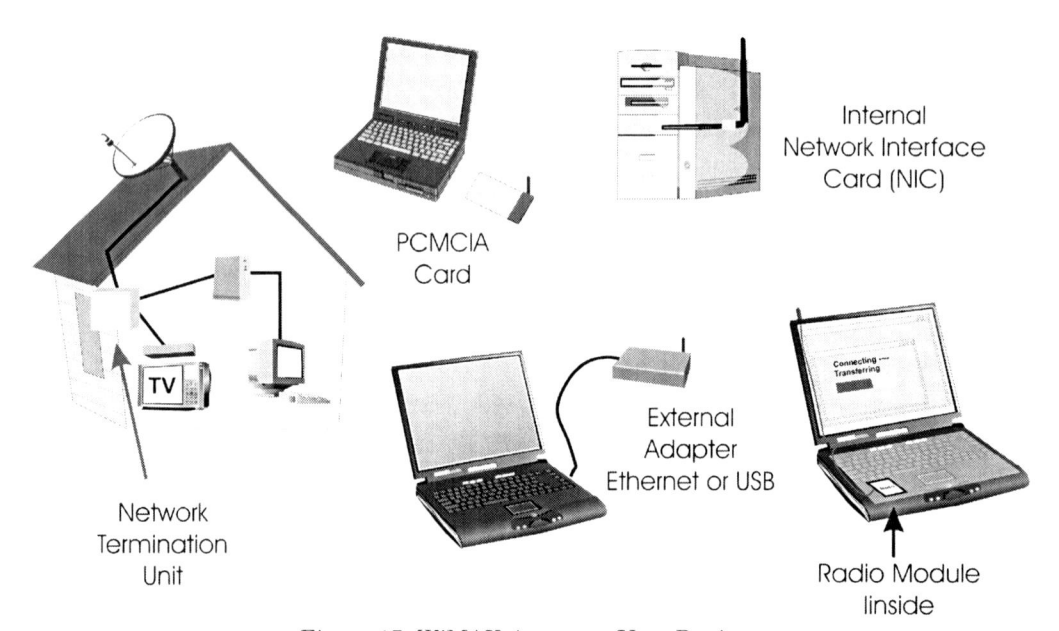

Figure 17: WiMAX Access or User Devices

WiMAX can offer service to multiple types of access devices. This figure shows several types of WiMAX user transceiver devices. For fixed service, a network termination unit communicates with the WiMAX base station. This type of unit is also called Customer Premise Equipment (CPE). Also available in areas where WiMAX provides indoor coverage to a PC is an internal *Network Interface Card* (NIC). Lastly, for portable and mobile use, a laptop can have an external Ethernet or USB WiMAX adapter, or ultimately an internal radio module.

Premises Distribution Equipment

A premises distribution network (PDN) is the equipment and software that is used to transfer data and other media in a customer's facility, home or personal area. A PDN is used to connect terminals (computers) and other media devices to each other and to wide area network connections. Some of the common types of PDNs are wired Ethernet, Wireless LAN, Powerline, Coaxial and Phoneline Data. Lucrative business models exist for operators that establish a PDN in a multi-dwelling unit (MDU), a multi-tenant unit (MTU) an apartment complex or an office park or campus.

Multimedia networking is the process of transferring multiple forms of media (e.g. digital audio, data and digital video) through a communication network (such as a home network). Multimedia networking is used to connect computers, media players and other media sources and players to each other and to send and receive media from wide area network connections. Care must be taken during the modeling of the business case to ensure there is enough capacity (spectrum) to deliver the multimedia envisioned.

The communication requirements for multimedia networks in the home are based on the applications that are used in the home along with how much and the times that the applications and services will be used. The typical types of applications that will be used in the home include telephone, Internet access, television, interactive video and media streaming.

The RFP responder should define which premises distribution networks they support, the versions of these systems and what forms of media they can transfer.

An RFP responder should define the types of access devices that are supported along with vendors and compatible model numbers.

Distribution Network Requirements

A WiMAX network consists fundamentally of terminal units (subscriber units, client devices, or customer premises equipment), and base stations (or access points). A *distribution system* connects the bases stations back to the *network operations center* (NOC) at the core of the network for concentration/aggregation and connection to the Internet, the PSTN, or Video Programming content providers through gateways to these other networks. A distribution system could be fiber, leased T-1 or DS-3 circuits, or point to point microwave connections, from base stations back to the NOC. These links haul the data traffic from the base stations back to the NOC, and are sometimes called *backhaul circuits,* or if extensive or if these links traverse a long distance, the *backbone network.*

A distribution plant is the physical property and facilities of a company (e.g. a telephone or cable TV company) that is used to distribute signals or services. WIMAX systems can use a mix of distribution types including copper, coaxial and fiber to connect to WiMAX base stations.

Network Architecture

Network architecture is the design, physical structure, functional organization, data formats, operational procedures, components, and configuration of a network. Network architectures usually divide network functions into layers of software and hardware.

Protocols

Protocols are the languages, processes and procedures that perform the functions used to send control messages and coordinate the transfer of data. Protocols define the format, timing, sequence, and error checking used on a network or computing system. While there may be several different protocol languages that can be used for communications services, the underlying processes (setup and disconnection of communication sessions) are fundamentally the same.

Data Capacity

Data capacity is the maximum data transmission or service carrying ability of a communications system. The unit of capacity measurement for the facility or system depends on the type of services or information content that are provided by the system.

Criticality

Some network operators have high criticality applications and services on their networks. For example, public safety, hospital, utility, and military organizations have routine as well as priority or high priority traffic on the network. The criticality of a particular application or service must be identified and the requirements clearly established.

Scalability

Scalability is the ability of a system to increase the number of users or amount of services it can provide without significant changes to the hardware or technology used.

Servers

A server is a computer that can receive process and respond to an end user's (client's) request for information or information processing. WIMAX systems use media servers for providing on demand content, distribution servers for routing signals to multiple locations, policy servers to establish and oversee quality of service levels (QoS), and other servers for distribution and network management functions.

The RFP responder should provide details on the proposed network architecture along with network evolution strategy. The data capacity capabilities should be defined along with system limits (e.g. number of households) and how the system can grow when limits are exceeded. The use of servers in the system should be defined along with their functions, such as distribution servers, network management changing servers and policy servers.

Testing Requirements

Testing capabilities are the systems and processes that can be used to confirm a product or service is operating correctly and within expected performance limits. Testing capabilities include system diagnostics and self-diagnostics.

Diagnostics are tests or programs (often built into a device or system) that test the functionality of the system and report the results. Diagnostic systems that are separate and simply monitor the operation of the subject system are considered to be non-intrusive.

Field Tools

Important for WiMAX testing are the tools needed in the field for testing and troubleshooting a network. These tools equip field engineers and technicians with the ability to diagnose problems and perform repairs. Without tools, operations and maintenance costs rise, and network reliability, maintainability and availability fall.

System Diagnostics

System diagnostics are the monitoring functions and processes that are used to identify the performance, operation and repair of equipment, protocols and/or transmission lines.

Self-Diagnostics

Self-diagnostics is a process and/or a program that is built into a device or network that allows users or devices to test the functionality of the system and report the results to assist in the finding of faults.

The RFP responder should describe the system test capabilities including system diagnostics and self-diagnostics.

Acceptance Testing and Commissioning

Acceptance testing is critical during the implementation stage, and therefore must be carefully set forth in the RFP technical requirements section. Testing is usually conducted by the vendor and witnessed by the network operator. Upon the successful completion of a series of tests the network owner / operator accepts the network from the vendor and commissioning takes place.

Commissioning is the release of a network from the vendor's program management and responsibility to that of the network owner. This is an important milestone, as it is the transition from one set of operations personnel to another. At this point the owner / network operator is operating and maintaining the network.

Authentication, Authorization, and Accounting (AAA)

An AAA system is an administration platform and process used in a communication network to authenticate users (determine that users are who they say they are), authorize users (allow certain users to have access to certain parts or services on the network), and account for user usage.

Conditional Access System (CAS)

A conditional access system is a security process that is used in a communication system (such as a broadcast television system) to limit the access of media to authorized users. Conditional access systems can use uniquely identifiable devices (sealed with serial numbers) and may use smart cards to store and access secret codes.

The RFP responder should define the suppliers of conditional access systems and what access devices (e.g. IP STB) the CAS system is compatible with.

System Administration

System administration is processes and tasks performed by a system administrator to add, change or disconnect (provisioning) devices, features and equipment. Administration may be performed directly or through remote locations.

Service Provisioning

Service provisioning is the process of an authorized agent or procedure that processes and submits the necessary information to enable the activation of a service. For WIMAX communication systems, this includes identifying services that can be offered, determining if these services can be provided at specific locations, sending and installing equipment to customers, and configuring equipment to deliver and receive services.

Remote Administration

Remote administration is the processes and tasks that are performed by a system administrator to add, change or disconnect devices, features and equipment from locations other than the general location of the system they are controlling. Remote system administration tasks may include monitoring the performance of the network and making configuration adjustments to the network as necessary.

The RFP responder should define the providing system and its local and remote management capabilities.

Customer Care

Customer care systems provide customer account tracking, service feature selection, billing rates, invoicing, and details. Customer care systems include call centers and customer self-care systems.

A call center is a place where calls are answered and originated, typically between a company and a customer. Call centers assist customers with requests for new service activation and help with product features and services. A call center usually has many stations for call center agents that communicate with customers. When call agents assist customers, they are typically called customer service representatives (CSRs).

Call centers use telephone systems that usually include sophisticated automatic call distribution (ACD) systems and computer telephone integration (CTI) systems. ACD systems route the incoming calls to the correct (qualified) customer service representative (CSR). CTI systems link the telephone calls to the accounting databases to allow the CSR to see the account history (usually producing a "screen-pop" of information). Call centers may be internal (direct) or outsourced. Call center outsourcing is the use of an outside firm to perform call center activities.

Customer self care is the process of allowing the customer to review and/or activate and disable services without the direct assistance of a customer service representative (CSR). Customer self care can be as simple as providing account billing information to the customer by telephone through the use of an interactive voice response (IVR) system to providing interactive service activation menus on an Internet web site.

RFP responders should define the expected requirements for customer care and the ability of the system to use internal and outsourced call centers. Customer self-care systems should be described along with how they relate to (reduce) call center activity.

Billing Systems

A billing system is a combination of software and hardware that receives transaction detail and service usage information, groups this information for specific accounts or customers, produces invoices, creates reports for management, and records (posts) payments made to customer accounts.

For WIMAX systems, billing systems need to have capabilities of paying complex sets of fees that include franchise fees, content licenses, royalties, sales commissions, and regulatory assessments. WIMAX systems may include the capability of processing orders and collecting fees for the sale of products and services other than data services.

Franchise fees are an amount charged or assigned to an account for the authorization to use a product, service, or asset. Franchise fees can be a combination of fixed fees or a percentage of sales. Content license fees are costs for the rights to use content. An example of a content license is permission to distribute a television program to users in a cable television system. Royalties are compensation for the assignment or use of intellectual property rights. Royalties may be paid on a time period basis or on a usage (e.g. number of times viewed) basis. Sales commissions are authorizations of compensation that are paid for the assistance with selling a product or service. Sales commissions may be a percentage of sales, a fixed amount or a combination of fixed fees and percentages. Regulatory fees are taxes, assessments or other amounts levied by regulatory authorities for the authorization to develop, operate or change systems or services that are overseen by the government.

Television commerce (t-commerce) order processing is the steps involved in selecting the products or services from a television catalog or advertising and agreeing to the terms that are required for a person or company to obtain products or services. T-commerce systems can be complex since they require interconnection with many types of systems, and with other companies.

A service detail record holds information related to a service or group of service usage events related to a communication session. This information usually contains the origination address of the user, time of day the service or services were requested, the types and quantities of services used, and charges that may be added from supporting vendors throughout the duration of the service.

Billing related events (such as a connection path or file transfer) are combined to create a billing record. Because billing events can come from many types of equipment and systems and can have various proprietary and standard formats, billing events are adapted by mediation devices. A mediation device is a network component in a telecommunications network that receives, processes, reformats, and sends information to other formats between network elements. Mediation devices are commonly used for billing and customer care systems as these devices can take non-standard proprietary information (such as proprietary digital call detail records) from switches and other network equipment and reformat them into messages billing systems can understand.

The RFP responder should provide details about the billing system, mediation devices and how the WIMAX billing systems will link and/or integrate with existing billing systems.

Reliability, Availability and Maintainability (RAM)

System reliability is the ability of a network or its equipment to perform within its normal operating parameters to provide a specific quality level of service. Reliability can be measured as a minimum performance rating over a specified interval of time. Also important is system availability, bit error rate, and mean time between equipment failures (MTBF). For communication systems, availability ranges from 99.9% to 99.999%, called "3-9s" or "5-9s" availability.

RFP responders should define their expected levels of reliability and redundancy along with how emergency response systems can identify, analyze and assist in the recovery of failed systems or services.

Single Point of Failure, Redundancy, and Battery Backup

Single point of failure analysis is typically performed on a system design to uncover all points in the network architecture that, if a failure occurs, would interrupt service to the entire network, or a significant portion of the subscribers or users of the network. This analysis typically reveals where in the network, *redundancy* or battery backup is needed. Redundancy is the designing-in of separate redundant parts of the network that allow for operation if one or the other parts are affected by failure. This design approach is costly, but required for high criticality applications.

System redundancy is a design of a system or network that includes additional equipment for the backup of key systems or components in the event of an equipment or system failure. While redundancy improves the overall reliability of a system, it also increases the number of equipment assemblies that are contained within a network. Redundancy usually increases cost.

Potential power outages lead to the need for battery backup. Batteries on the subscriber units enable the subscribers or users to operate the network despite a power outage. The requirement for battery backup can be from five to ten minutes, the typical outage time for commercial power interruptions in the United States, or it can be up to 8 hours or even 24 hours for public service, utilities or the military, where network operation is essential, during potential outages of longer duration.

Disaster Recovery

Disaster recovery is the processes that are used to restore services after a significant interruption (disaster) in communications systems. Disaster recovery processes usually occur after events such as fires, floods or earthquakes. However, disaster recovery may also occur after critical equipment failures or information corruption that occurs from software viruses. Disaster recovery ability may be defined through the use of system reliability and redundancy criteria.

The RFP Development Process

The RFP development process is tasks that are performed to discover and define the needs of the company that can be used to produce an RFP document.

The RFP development process starts with the company recognizing and defining the needs (such as WIMAX services) that the RFP will cover. The person or approving authority is identified from the company who can issue and approve the RFP. Using information gathered from multiple departments and potentially gathering information from pre-approval conferences, the RFP requirements are defined. A list of potential vendors is created (RFP invite list) and the RFP is released. RFP participants may submit questions or ask for clarifications on key areas of the RFP. Clarifying releases and responses to clarification requests may be issued. RFP participants then submit their proposals for evaluation. A committee usually reviews the proposals and selects a winner. The winner then negotiates the details of the contract and a purchase order is issued so work can begin.

Figure 18 shows the general process used for the initiation, development and completion of the RFP process. This diagram shows that the process generally begins with the RFP issuer and the RFP responder does not become involved until they are invited to attend an RFP conference. The RFP issuer spends a substantial amount of time and effort in determining the requirements that will be defined in the RFP document. This process shows that the RFP creation process is typically interactive between the issuer and responder with questions, responses and clarification of information being exchanged during the RFP process. This diagram shows the RFP process typically ends with an announcement of the vendor (or vendors) that has been selected as a winner of the RFP.

Creating WiMAX RFPs

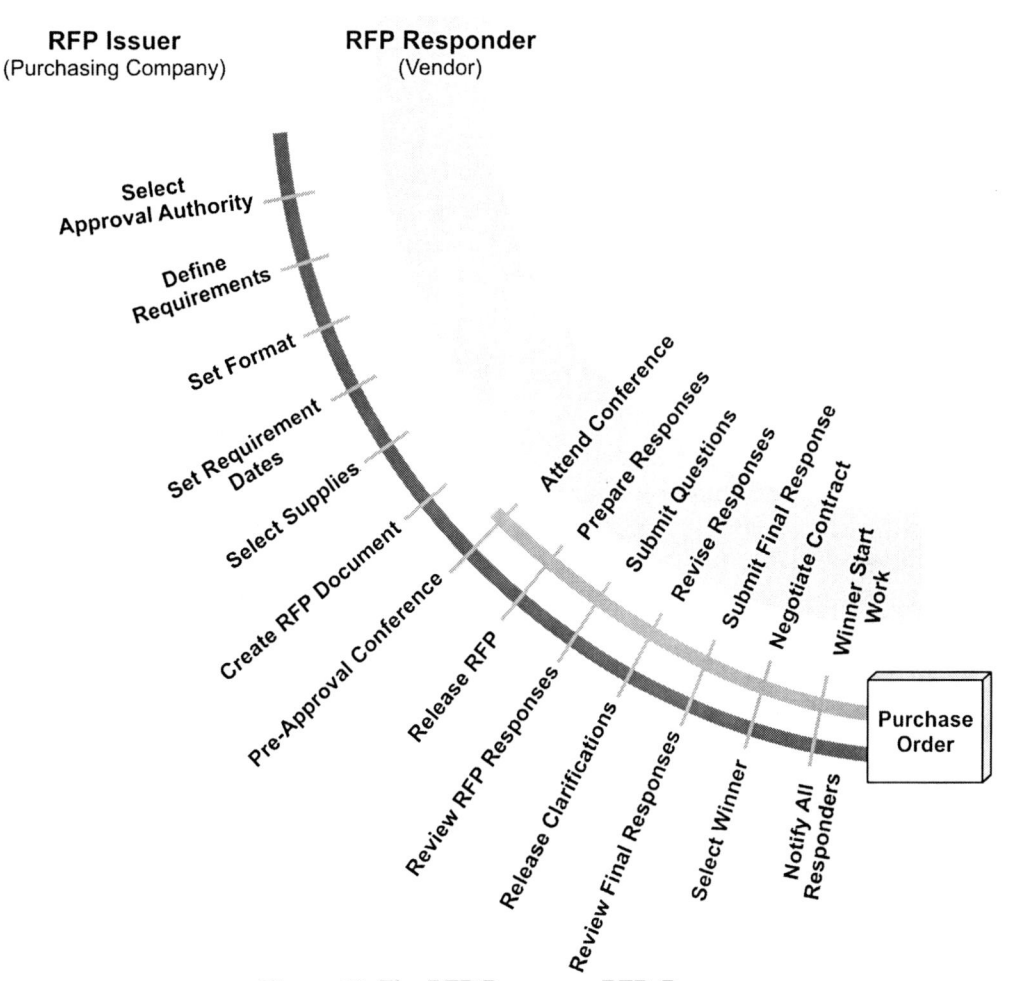

Figure 18: The RFP Processag_RFP_Process

Qualifications

Qualifications are indicators that a vendor has the capabilities to provide the products or services that they define in their RFP response. An RFQ and RFI can precede an RFP process to determine qualifications, but this determination can also be done, or confirmed in an RFP. RFPs may ask companies to provide information on their financial, technical, and operational experience, in addition to any licenses and certifications.

Financial Qualifications

Financial qualifications are materials or information that can validate the ability of a person or company to perform projects or services. Financial information is data that represents the financial performance of a company or person. Financial information may include income statements, balance sheets, cash flow statements, and estimates on how these financial indicators may change over the next few years (financial projections).

Technical Qualifications

Technical qualifications are the skill sets and experience that a person or company has that enable them to perform certain types of tasks or projects. Technical qualifications can be indicated by the types of workers, training programs attended and certifications that staff have been awarded.

Operating Information

Operating information is data that relates the ability of a business to provide products or services. Operating information may include production capabilities, previous project management experience and key business relationships.

Licenses

A licensee is the holder of a license that permits the user to operate a product or use a service. In telecommunications, a licensee is usually the company or person who has been given permission to provide or use a specific type of communications service within a geographic area. To perform certain types of tasks or projects, trade or professional licenses may be required.

Certifications

Certifications are statements of qualification for individuals, groups or businesses. Certifications may be required to work with certain types of products or systems. Companies may complete self-certification statements (such as an implementation conformance statement) to identify they are qualified and have a specific set of capabilities.

Supporting Vendors

Supporting vendors are companies that may be used to assist an RFP responder to provide products and services. If supporting vendors are used, their roles and qualifications should be provided in the RFP response.

Existing System Evaluation

Existing system evaluation is the review of the current systems and processes to identify and define existing equipment (if any) that may be used, integrated or upgraded with a new WIMAX system and services. Existing system evaluations should include the facilities, their locations, equipment and software applications. Key details should be provided on the distribution plants, access systems and back office systems.

Distribution Plant

A distribution plant is the physical property and facilities of a company (e.g. a telephone or cable TV company) that will be used to distribute signals or services within a building or campus. A key asset to operators may be the right of way and easements that allow for the installation and maintenance of communication lines along public and private (e.g. commercial buildings) areas.

Access Systems

An access system is a portion of a communication system that coordinates requests for services and enables the transfer of information when authorized. The types of access systems in use for customers (such as DSL, cable modem or wireless broadband) should be designed to help determine the best access equipment and premises distribution options.

Back Office Systems

Back office operations are the processes and systems that are used to assist with the operation and management of communication systems. Back office operations usually include billing and customer care systems, accounting, maintenance services, and asset management. It is important to define the capabilities of the back office systems including network interface options, platforms used and performance capabilities.

Needs Assessment

A needs assessment is a review of the human, equipment and operational resources that are necessary to achieve future objectives. A needs assessment may be composed of several steps including visioning sessions, information gathering, systems evaluations, and public hearings.

The RFP process involves obtaining information from multiple departments within the company about their specific needs and objectives. The RFP information gathering process will assist a company in understanding what needs to be placed inside a request for a proposal, how to issue a request for a proposal, and dealing with issues relative to responding to RFPs that have been received from supporting vendors.

Advisory Committee

An advisory committee is a group of people who review and provide information that helps a company and/or person to make decisions.

Visioning Session

A visioning session is a meeting between two or more participants with the purpose of reviewing and/or defining future objectives and the likely steps or processes that are necessary to achieve these objectives.

Figure 19 shows the general process that is used to determine the requirements to be defined by the RFP. This diagram shows the process typically starts with an executive decision to initiate a project to install or upgrade a communications system. The purchasing department may request information and organize meetings from departments to determine the needs of each department. These departments usually include information technology (IT), operations (Ops), sales and marketing, engineering, accounting, and executive departments. These requirements are combined (merged) into the requested requirements that are reviewed and approved by the company executives (or a committee created by the executives).

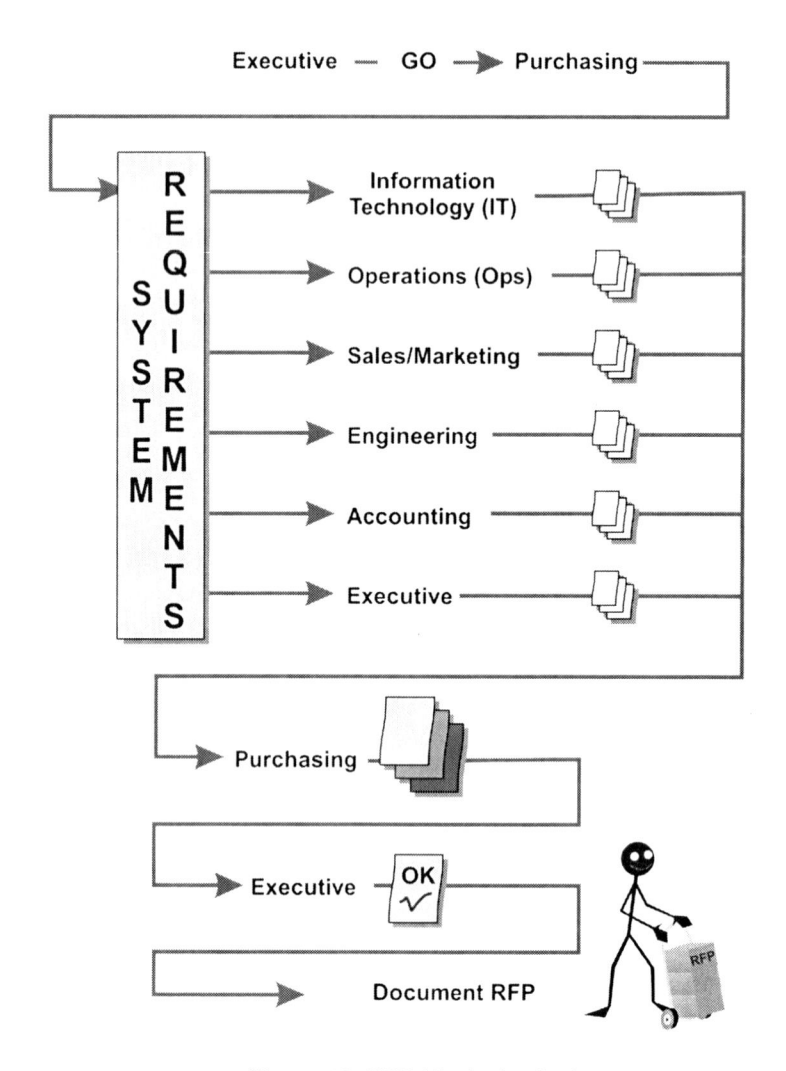

Figure 19: RFP Needs Analysis

Issuing, Approval and Contract Authority

The issuing and approving authority is the agent within a company that has the authority to issue the RFP and sign off on any contract stemming from the RFP process. The approval authority may be the CEO, president of the company or it could be a purchasing or project manager. There may be multiple approval authorities for different portions or requirement areas in the RFP. The RFP should document the following information of the authority:

Name
Position
Street Address
City, State and ZIP Code
Telephone Number
Fax Number
Email Address

The issuing authority may be required to sign off on all pertinent requirements that are defined by the RFP development team. This can help to ensure that the RFP contains the necessary requirements for each of the issuing companies' divisions.

Public Notice

Public notice is a message that is communicated to the pubic in general (usually through advertising or public media channels) that alerts interested parties that a change in regulation is occurring. The issuance of some RFPs (such as government agencies or schools) may require public notice for the issuance of an RFP.

Contracting Authority

The contracting authority for an RFP is the person or organization within the company that has the authority to offer a contract to the successful responder to the RFP. Because a WiMAX communication system contract can impact multiple departments within a company, it is important to identify the approving authority.

RFP Project Dates

RFP requirement dates are the dates by which specific actions must be taken to successfully participate in an RFP process. Failure to complete a task by one or more requirement dates typically disqualifies the vendor from continuing in the RFP bidding process. There are several key dates that may be defined in an RFP ranging from the date of submission to the notification of the RFP winner and all the other RFP respondents. The documented step could include:

Conferences

A pre-proposal or bidder's conference is a meeting where the requirements for a proposal document are described and discussed, and questions are answered. These conferences can come before issuance, just after issuance, or just prior to the closing of an RFP submission window. Advantages of early conferences allow for complete introduction of the RFP. Later conferences allow respondents to internalize the RFP and come prepared with good questions.

Clarification Requests

Clarification requests are questions that ask for details or an expanded description of terms in RFPs or other documents.

Clarification Responses

Clarification responses are details or expanded descriptions that are provided to people or companies that sent clarification requests for RFPs or other documents. Typically, clarification requests from all respondents are combined and issued in one batch to all respondents at one time, thereby precluding one respondent from receiving information that is different from, or more than, information provided to another respondent.

Pre-Response Conferences

A pre-response conference is a meeting towards the end of an RFP response window where additional details about a request for a proposal document are described and discussed. A pre-response conference may be held as a result of the receipt of many clarification requests on specific items which indicates the information regarding that material in the RFP is not clear to the RFP responders.

Site Review

Site review is an inspection of a location to determine existing and/or upgradeable capabilities of devices, systems or facilities.

RFP Release Date

The RFP release date is the date that an RFP is shipped or made available to participating companies.

Closing RFP Submission Date

A closing submission date is the last date possible to submit a document or proposal (such as an RFP).

RFP Award Date

An award date is the day that a contract or project is authorized or a legal commitment occurs.

Contract Negotiation Date

The contract negotiation date is day that negotiations begin to define the details of a project or contract.

Final Signing

Final signing is the day the contract to provide products and services defined by the RFP becomes binding on the RFP responder.

Figure 20 shows the general timeline of events that generally occur in a WiMAX RFP process. This timeline shows that the first steps involve the determination and approval of the company's WIMAX system and service requirements. Vendors are selected by invitation into the RFP process and a pre-release (a week or so prior to RFP release) and/or pre-submission (a week or so prior to RFP submission) conference may be held. When the RFP is released, there is generally a time period that is allowed for the vendors to review and respond to the questions in the RFP. During this time period, clarification questions may be received and responded to, typically on an automated website service. At a specific date, the winner of the RFP is selected. Both winners and losers are notified. The winner enters into a contract negotiation period and the both parties become legally bound on final signing day.

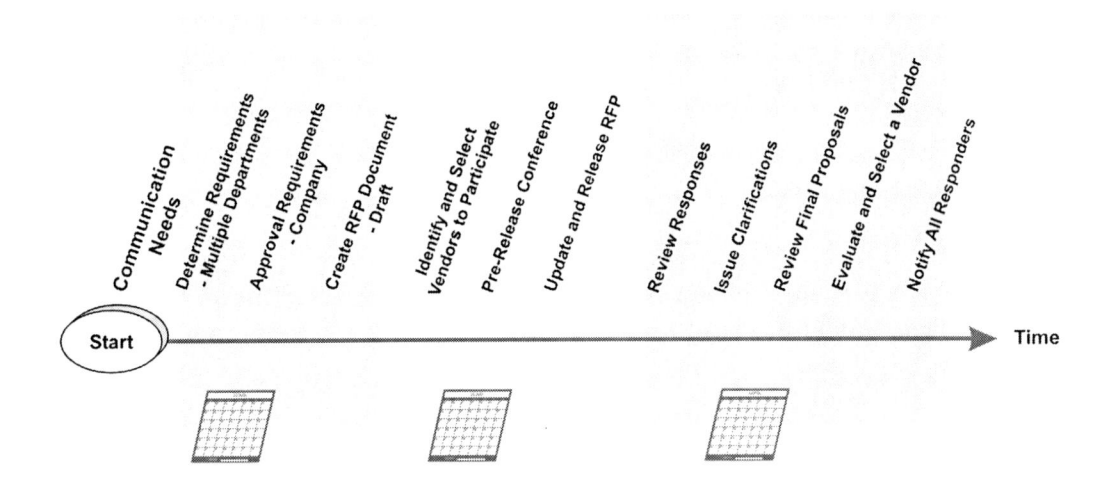

Figure 20: Sample RFP Timeline

Implementation Plan

Implementation requirements are a description of how the responding company must manage the implementation (installation and setup) of the WIMAX system. This description should include the roles of key project personnel and the winning bidder should be required to present a detailed implementation plan. The implementation plan should include an installation schedule, initial operational capabilities, alpha testing, beta field testing, full operational capability, system cutover, and final acceptance testing.

An installation schedule is the sequence of time events that will occur for the delivery and installation of products or services. The installation schedule should consider information that could impact the ability to obtain and install equipment such as lead times and access to facilities.

Initial Operational Capability (IOC) Date

The initial operational capability date is the time when a system or service is initially activated. The responder should define the initial operational capabilities of the system and any phases of new system capabilities that will be provided.

Alpha Testing Date

Alpha testing is the first stage in testing a new hardware or software product, usually performed by the in-house developers or programmers. Alpha testing is the initial internal and possibly limited field testing process used to confirm the operation and performance of new hardware or software products. The key purpose of Alpha testing is to identify basic problems during typical operating conditions. The typical number of Alpha test participants is 10 to 50.

Field Trial Date (Beta Testing)

Field trial date(s) are the times that tests will be conducted by users of a product or service in their native (field) environment. Field trials are used to determine if the services or systems are operating or performing to specification levels that have been set by the buyer of the system or services. Field trials are also called beta tests.

Beta testing is the field testing process used to confirm the operation and performance of new hardware or software products before a product is officially released. Beta testing is usually performed by friendly customers or affiliates of the manufacturer or developer. The key purpose of Beta testing is to identify problems and the reliability of operation during normal field operating conditions. The typical number of Beta test participants is 50 to several hundred.

Full Operational Capability (FOC)

Full operational capability is the ability of a system to provide all of its defined services within the defined operational limitations of the system.

System Cutover

System cutover is the process or date where new systems or equipment are used to provide services to customers (old systems are cut off).

Acceptance Testing

Acceptance test is a test that evaluates the successful operation and/or performance of an electronic assembly or communication system. Acceptance tests usually have specific operation requirements and test measurements. Acceptance tests are often used as a final product approval and may authorize a product for production or purchase.

RFP Response Procedure

An RFP response procedure is the instructions to the RFP responder that defines the steps that the responder must perform to comply with the terms of the RFP submission process. The response procedure may define levels of compliance and how variations from compliance will be handled. Levels of compliance to the specifications of the RFP may include, for definitive requirements:

- Comply (i.e. responder meets the definitive requirement and provides technical specifications that document compliance)
- Partially Comply (i.e. complies with part of the requirements. The vendor must provide an explanation of those requirements met, and those not met, and the impact on technical specifications)
- Comply with qualification (i.e. meets requirements in different manner and additional details are needed)
- Exceptions (i.e. response does not meet requirement and the responder needs to supply an alternative to continue)

For probing technical requirements the respondent must:

- Comply (i.e. responder meets the intent of the probing requirement, restates the requirement in vendor specific terms and provides technical specifications that document compliance)
- Partially Comply (i.e. complies with part of the requirements. The vendor must provide an explanation of those requirements met, and those not met, and the impact on technical specifications)
- Comply with qualification (i.e. meets requirements in different manner and additional details are needed).
- Exceptions (i.e. response does not meet requirement and the responder needs to supply an alternative to continue)

RFP Issuance

RFP issuance is the copying and distributing of an RFP to a list of authorized recipients. This distribution may be in the form of paper and/or electronic copies. Potential recipients of RFPs may have received and responded to an RFP invite and for some RFP issues, only vendors that have been invited to participate may respond. The issuance of an RFP may also involve the signing of non-disclosure to ensure that confidential information about the company issuing the RFP (such as new service and expansion plans) is not made available to the public or to competitors.

Prospective Suppliers

Prospective suppliers are a group of companies or people who have characteristics that indicate that they are likely to offer or they are qualified to produce certain types of products or services.

Finding prospective suppliers involves identifying companies with experience and characteristics that are more likely to ensure the RFP winner will be able to provide the systems, products and services that are defined in the RFP. Finding prospective suppliers can be started by contacting industry associations, trade magazines and reviewing industry directories. Once an initial list is created, related companies can be identified by searching for related companies (related links) on www.Alexa.com (a web site popularity ranking tool). Additional companies may be identified along with gathering preliminary details from interviews with companies that purchase similar (non-competing) companies.

Companies that are identified as a potential RFP participants may be sent an RFP invite letter or package. An RFP invite is a request from an RFP issuer that is sent to a company to invite them to participate in an RFP process. An RFP invite may define the products and/or services desired along with some of the basic terms (e.g. non-disclosure agreement) and/or qualifications (company size, technologies and support capability) for a company to participate in the RFP process.

In the event additional potential RFP respondents are identified after the RFP is released, they should only be added to the list if they can respond on the same information and follow the same guidelines and schedules as all the other RFP participants.

Before issuing an RFP, it may be reviewed by the legal department to ensure it complies with legal guidelines. It is also likely that RFPs will be reviewed by other departments within a company that may be affected by the RFP. The departmental review of RFPs also helps to ensure that department leaders will be more helpful when implementing the RFP as they may feel they participated in its creation.

Some of the departments that are involved in the preparation of an RFP include purchasing, legal, accounting, and public relations. The purchasing department influences the approval requirements and the formatting of the RFP. The legal department ensures key terms and conditions are included and well defined. The accounting department defines the financing requirements. The public relations department can contribute company information. The operations department can provide desired implementation milestones and timelines.

Figure 21 shows how the creation of a request for proposal (RFP) document typically involves contents and feedback from multiple departments within a company. Approval and formatting requirements come from the purchasing department, terms and conditions from the legal department, financing terms from the accounting department, company background information from the public relations (PR) department, and critical date requirements from the operations (Ops) department.

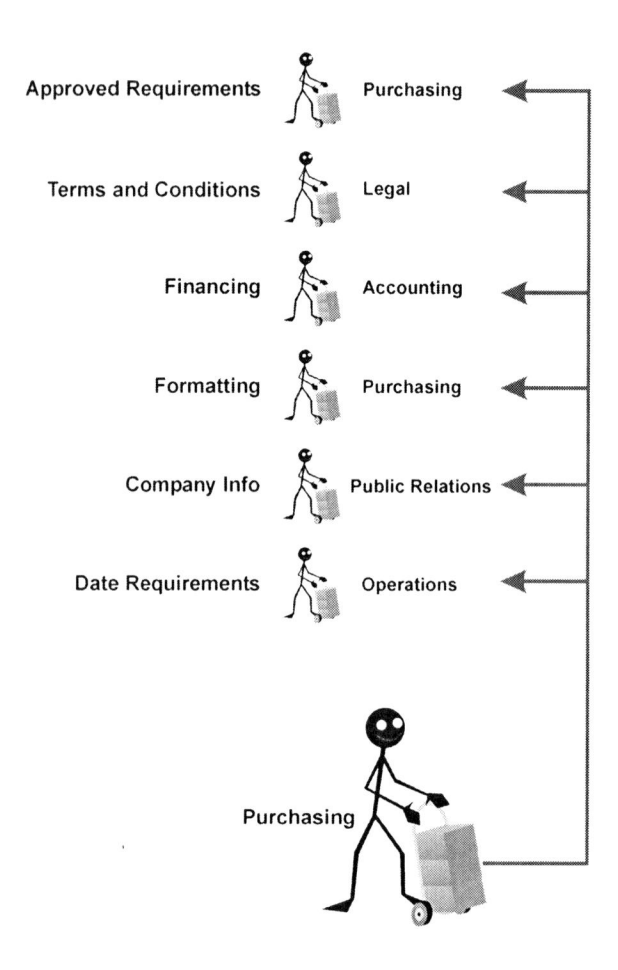

Figure 21: RFP Document Creation Process

RFP Response Requirements

RFP response requirements are the specific set of actions, documents and data that a company must perform and describe when submitting an RFP response. The RFP requirements may include information types, a timeline of when the information is to be provided, the required format of the information provided (printed and or electronic documents), and additional details about actions that may occur between the RFP issuance and response processes.

Some respondents prefer to bid both individually and through others, such as partners, joint ventures, system integrators or prime contractors. The issuer must decide if this is allowable and so inform the respondents in the RFP.

A useful item in the RFP package is a checklist of items required in the submittal package. This could include:

   The RFP document with responses imbedded in the document
   The financial spreadsheet form with cost information included
   The signed legal terms and conditions addendum
   Commercial brochures on the company products
   Corporate profile form with information on the respondent

RFP Approval Process

An RFP approval process is the steps that are taken from the reception of RFP responses to selecting a winner of the RFP. The processes used to evaluate and approve may be included in the RFP to ensure the communication and responses to the RFP are handled by the proper authorities within the issuing company.

Pre-Proposal Review

A pre-proposal review is a meeting where the anticipated requirements for a request proposal document are described and discussed. A pre-proposal review can be an informal environment that allows for explanations and discussion of what is needed and some of the available options that may meet those needs. A pre-proposal review can help to shape and clarify an RFP before it is released.

Releasing RFPs

Releasing an RFP is the process of distributing the documents and or electronic files to the RFP participant list. When an RFP is released, electronic access to RFP participants may be provided that allow them to obtain the RFP and supporting documents via a web site. Each RFP recipient should acknowledge their acceptance and receipt of the RFP documents.

RFP Responses

Request for proposal (RFP) responses are documents and supporting materials that are provided by companies that have received and reviewed an RFP that define how it can satisfy the objectives and requirements defined in the RFP. When RFP responses are received from vendors, the responses are logged in, reviewed and marked with an identifying code.

RFP Response Review

RFP response review is the process of receiving, reviewing and evaluating responses to RFPs. As RFPs are received from responders, they are marked and/or logged as received with the date of receipt. To ensure that RFP participants have filed their responses within the required submission due dates, a process of logging in the responses should be used.

Clarification Questions

Clarification questions and requests may be received before the RFP response date. The receipt of clarification questions or requests should be logged and documented.

The RFP responder may be informed that contacting people within the company regarding the RFP that are not on the authorized contact list is not allowed. Contacting the company regarding the RFP through employees other than those specified by the RFP could potentially influence the contents of the RFP or disqualify the RFP responder.

Clarification Response Review

Clarification response review is the interpretation of clarification requests and the determination of what actions and expertise are appropriate for the request.

Issuing Clarification Updates

Clarification updates are messages or documents that are sent to RFP respondents. If a clarification request is of a general nature, a clarification update may be issued to all RFP participants.

Response Evaluation

RFP response evaluation is the process of evaluating and judging the RFP responses. RFP response evaluation is usually based on a pre-defined set of criteria.

Evaluation Criteria

Evaluation criteria are the determination and weighting of criteria. Evaluation criteria may include the vendor's previous experience with the systems they are proposing, their financial and business stability, and how well their proposal satisfies the requirements of the RFP.

Figure 22 shows some of the criteria that may be used to select a vendor who responds to a WiMAX RFP. The criteria typically include the vendor's business assets, experience, radio and data networking capabilities, and implementation wherewithal. The financial qualifications of the company may include profitability (profit and loss statements), liquid and fixed assets (balance sheet) and other performance factors (sales in the product area). The vendor's technical experience may include project development, and industry participation in developing related products and services. The vendor's radio and data networking technology qualifications may include experience in deploying similar applications (voice, video, and data), system types (licensed and unlicensed spectrum, fixed, mobile, outdoor and indoor systems), IT and security capability, employee skill sets, patents and soft-

Supplier Selection Criteria	Score
Business	
Annual Revenue	☐
Networks sold	☐
Number of Employees	☐
Experience	
Data Applications	☐
Voice Applications	☐
Video Applications	☐
System Integration	☐
System Implementation	☐
Radio Networking	
Unlicensed Operation	☐
Interference Mitigation	☐
Mobility Systems	☐
Rural Systems	☐
In-building Operation	☐
Data Networking	
Security	☐
IP Architecture	☐
Operational Support Systems	☐
Business Support Systems	☐
Implementation	
Training	☐
Support	☐
Program Management	☐
System Integration	☐

Figure 22: Supplier Selection Criteria

ware programs. Lastly, the vendor's support through implementation and initial operations is critical for a smooth commissioning of the network into operations.

Decision Matrix

A decision matrix is a table or related set of data that contains qualifying rules and quantifying processes (rating) of criteria that is used to make a decision (such as selecting a vendor to supply a product or service). A decision matrix can be used to simplify and remove bias from the evaluation of RFP responses. A decision matrix could be expanded by the addition of appropriate columns for score, weighting, total and grand total for each respondent.

Criteria Weighting

Criteria weighting is the application of scaling factors to criteria that reflect their level of importance. Not all criteria have the same importance level and criteria weighting can adjust for importance. For example, experience in a particular industry or technology might be more important than the financial strength of the company.

Scoring

Scoring can be via individuals with scoring criteria and weightings based on importance of the criteria to the program. The scoring can be done in groups to allow for normalization of expectations among group members. And scoring can be a step by step gating function whereby vendors that make a technical gate, proceed to a risk gate, and then finally to a cost ranking. The exact method of scoring depends on the degree of accuracy needed by the organization. Sometimes high accuracy and objectivity is required when the selection is political, emotionally charged, or the organization is influenced by a fastidious consensus driven, decision making process.

Winner Selection

A winning responder is a person or company that has been selected as a vendor or provider of services as the result of the proposals or responses that they submitted upon request for proposal. A standby winner may be identified in the event that the winning bidder is unwilling or unable to enter into a contract.

Figure 23 shows a general process that may be used to select the winner of an RFP. Each eligible vendor must provide a timely, complete, and correctly structured response to the RFP; the bidder should be compliant with the requirements of the RFP as well as demonstrate a low risk feasible solution. Innovation is important when designing a solution to a complex set of requirements, and is usually rewarded. The bidder should be in agreement with all standard legal and procurement terms and conditions for system purchases. The responses to the financial sections should be correctly structured and complete. And after coordination and careful judgment the stakeholders in the organization that will own the system should all concur for the most part, with the winning selection and the multiple divisions within a company may be required to accept (sign off on) the vendor's responses to the RFP questions.

Vendor Selection	Check of
RFP Submission	
Timely submission	☐
Packaged correctly	☐
Package complete	☐
RFP Compliance	
Compliance with mandatory requirements	☐
Technical risk	☐
Feasibility	☐
Innovation	☐
Terms and Conditions Compliance	
Legal signoff	☐
Procurement signoff	☐
Financial	
Bid structured correctly	☐
Bid complete	☐
Coordination	
Operations concurrence	☐
Engineering concurrence	☐
Regulatory concurrence	☐
Executive concurrence	☐

Figure 23: Vendor Selection Process.

Short-Listed Vendor Meetings

The RFP process up to this point is essentially an *arms-length, paper-process*, meaning vendors are formally treated as a group and given the opportunity to ask questions and receive formal answers. There may be no equipment demonstrations or vendor meetings up to this point. This ensures all vendors are treated equally, and the issuer has the best chance of receiving information that can be compared in *apples-to-apples* (fairly and equally) fashion. What might be needed after a scoring process and a prospective winner and runner(s)-up is established, is the invitation of vendors to a meeting to address any risks, issues, questions, or concerns that might be remaining after the paper ranking process. This is the last opportunity to modify the scoring, and potentially re-order the ranking of the list of vendors.

Notification of Award

Notification of an award is the informing of a person or company who submitted a proposal or bid that their offer or part of their offer has been accepted.

Notifying of Rejection

Notification of a rejected offer is the informing of a person or company who submitted a proposal or bid that their offer or part of their offer has not been accepted.

It is important to maintain a good relationship with the runner(s)-up of the RFP process. These competitors may be available later for a trial or pilot bake-off, multi-vendor environment *second-source* vendor, or just as an option if the primary vendor does not measure up or meet expectations.

Protest of Award

Protest of award is the process of requesting that an RFP issuer review an award selection for various causes such as asserted biased evaluation, errors and omissions or other criteria with basis. The process for protesting awards may be defined in an RFP and/or procedures may be defined by government regulations (such as when a contract is awarded from a government). The original RFP may define a time period within which protest of awards must be filed after an award selection has been made.

After a protest of award is received, a stay of procurement may be issued. A stay of procurement is a delay in the issuance of an award or suspension of purchases from a winning vendor until an issue (such as a protest of an award) is resolved.

RFP Contents

This section provides an outline for a WiMAX RFP. The outline is followed by an explanation of each of the main items below. Also, refer to the table of contents and index to find information that should be included in the elements of this outline.

RFP for a WiMAX Network Template

I. Summary Needs and Purpose Statement
    a. Executive summary of scope and interest for the RFP
    b. Purpose statement on the intent of the RFP process

II. General Procedures
    a. RFP Submission Procedure
    b. RFP Transmittal Process

III. Issuing Company Information
- a. RFP requirements summary
- b. Company background
- c. Existing systems and services
- d. RFP objectives
- e. Scope of work
- f. Instruction to respondents
- g. RFP distribution
- h. Bidders' conference
- i. RFPs clarifications
- j. Responses requirements
- k. Implementation schedule
- l. Subscriber forecasts

IV. Respondent Information
- a. Company background
- b. Qualifications
- c. Experience
- d. Respondent's references
- e. Respondent's support capabilities
- f. Supporting vendors and credentials

V. WiMAX Requirements
- a. Business Requirements
- b. Technical Requirements
- c. Coverage Requirements
- d. Capacity Requirements
- e. Frequency Plan/Interference Mitigation Requirements
- f. Broadband Data Access
- g. Internet Access
- h. Voice and Telephony
- i. Video and Television Services
- j. Capacity Requirements
- k. Frequency Plan & Interference
- l. Distribution System

- m. Access Devices
- n. Security
- o. Technology Refresh and Roadmap
- p. Testing
- q. AAA and CAS
- r. Business Support and Billing System
- s. System Administration and Operational Support
- t. Maintenance, Spares and Repair Parts
- u. Customer Care
- v. Disaster Recovery

VI. Implementation Schedule

- a. Initial Operational Capability
- b. Alpha Testing
- c. Field (Beta) Testing
- d. Full Operational Capability
- e. System Cutover
- f. Acceptance Testing
- g. Training requirements
- h. Pilot, Proof of Concept, or Trial

VII. Procurement Terms and Conditions

- a. Terms and Conditions
- b. Change Orders
- c. Compliance
- d. Nonresponsive proposals
- e. Performance guarantees
- f. Proprietary information
- g. Warranty
- h. Liability
- i. Regulatory compliance
- j. Right to reject

VIII.   Pricing and Financing Options

    a.   Equipment pricing
    b.   Support services
    c.   Financing Terms

An RFP typically includes a general procedure section, issuing company overview, responding company information request, requirements and specifications, RFP terms & conditions, pricing & financing formats, and guarantees.

Summary of Needs and Purpose Statement

The typical WiMAX RFP provides an executive summary of the contents of the RFP. An intent or purpose section provides further detail of the expected outcome of the RFP program.

General Procedures

General procedures inform the respondents of the general processes to follow during the RFP program. It can include timelines, response instructions, forms to submit, and other compliance statements

RFP Submission Procedure

The RFP submission procedure is the sequence of tasks, delivery components and dates that a company or person must use to submit a response to an RFP. The submission procedure section of an RFP should include the method of response and how interactions between an RFP responder and RFP issuer should occur. The submission procedures usually include details on the expected preparation and format of the RFP response, the methods the company will use for receiving responses to the RFP, how RFP responses will be evaluated, and how RFP respondents will be notified of the winner of the RFP.

RFP Transmittal Process

RFP transmittal is the process used to submit an RFP. RFP transmittal requirements include who the RFP is submitted to, when it should be received, the contents of an RFP, supporting documents, and the format. RFP submittal formats may include electronic and/or printed copies. Third party services [6] have emerged that help organize and manage the process of RFP issuance, response, and data management.

Issuing Company Information

Issuing company information contains a description of the business and operations of the company that is issuing the RFP. The company information may include company background, business objectives, products and services offered, and operational capabilities.

RFP Requirements Summary

This is a high level summary of the requirements, both business and technical, that appear in later sections.

Company Background

Company background is a description of the key attributes and objectives of the issuer company. Issuer company background may include product types and industry categories a company participates in, along with the core values of the company. To specify company requirements for any respondent to the RFP, it is helpful to have an executive summary of the company's background.

Existing Systems and Services

Here a company may describe existing, or legacy networks or systems, particularly ones that the new system may interconnect to, or replace.

RFP Objectives

The objectives of the RFP are clearly and succinctly listed here.

Scope of Work

The scope of work (SOW) is referenced here (e.g., See section V. WiMAX System Requirements, etc.).

Instruction to Respondents

This section provides details to the respondents

RFP Distribution

This section describes the method by which the RFP will be distributed to respondents.

Bidders Conference

This section provides time, date, place and instructions regarding the bidders' conference.

RFPs Clarification

This section provides instructions on how to submit questions and the method of receipt of answers to the RFP.

Response Requirements

This section provides instructions on how to submit a response.

Respondent Information

Respondent Company Information Request

Responding company information requests contain the types of company information that the RFP issuer desires to know about the RFP responder. The responding company information may include company background, products and services offered, development capabilities, experience in the industry, related client history, and financial information.

WiMAX Network Requirements

Business Requirement

Business requirements contain general requirements of the proposed business, stated in a manner that allows the respondent to reply with information specific to the respondent's products and proposed solution. Requirements may be stated as definitive or probing.

Technical Requirements

Technical requirements are needs of the issuer that describe at a technical, operational (high) level, the essential functional characteristics of products, materials or services needed. Requirements should be stated (written) in a manner that allows operational testing to confirm whether or not a respondent has met the requirement.

The respondents take these requirements, and through their individual design process, with their individual products, materials or services, produce detailed technical specifications that describe clearly all technical attributes of the product, material or service. These specifications should be stated by the vendor in the response in a manner that allows, if necessary, detailed testing of the technical specification.

Industry standards can be attached as technical requirements or technical specifications. Industry standards are protocol, software, or hardware specifications that allow multiple vendors' products to interoperate. They are typically created through the participation of multiple companies that are part of a professional association, government agency or private group.

The network technical requirements of a WiMAX RFP typically include all the important factors of a complete network. These factors include access, distribution, devices, data, voice, video, testing, AAA / CAS, billing, system administration, and disaster recovery.

Implementation Schedule

Milestones in an implementation schedule for a major WiMAX network deployment include the following:

- Subsystem and development capability and test
- Initial Operational Capability
- Alpha test
- Field (Beta) test
- Full Operational Capability
- System Cutover
- Acceptance test
- Commission to operations

Testing Requirements

Testing is a key activity during implementation, and its description in the RFP allows respondents to plan, scope, and budget for this important phase of the program.

Testing requirements are the list of tests that must be performed for the acceptance of a product or service. Testing requirements usually include functional tests, performance tests and stress tests. Functional tests are observations and/or measurements that are performed during normal operating conditions of a device, service or system to determine if it can perform its designed functions. Performance tests are measurements of operational parameters during specific modes of operation. Performance tests are used to determine if the device or service is operating within its designed operational parameters. Performance tests can be performed over time to determine if a system is developing operational problems. Stress tests are observations and/or measurements of devices or services under operational conditions that are near or above their design limitations. Stress tests are performed to determine how a network or system will operate under loaded or congested conditions.

Figure 24 shows some of the many items that may be required for the acceptance of a system that is covered by an RFP. This list shows that the acceptance tests typically include functional and performance testing of services and features under normal and emergency conditions.

System Acceptance Tests	Pass/Fail
Applications	
Data Applications	☐
Voice Applications	☐
Video Applications	☐
Devices	
Fixed CPE	☐
Mobile Terminal	☐
Portable radio NIC	☐
Sensor	☐
Actuator	☐
Coverage	
Percent area coverage	☐
Percent population coverage	☐
Percent in-building coverage	☐
Mobility and handoff	☐
Roaming	☐
Security	
Intrusion	☐
Encryption & Privacy	☐
Authentication & Authorization	☐
Compartmentalization	☐
Implementation, Operations & Maintenance	
Program Management	☐
System Integration	☐
Network Integration	☐
Training	☐
Support	☐
Network Management	☐
Training	☐
Warranty, Spares, Repair parts	☐

Figure 24: System Acceptance Criteria.

Training

Training of personnel that will assume operations and maintenance duties of a WiMAX network is important. Training requirements are skill set development objectives that are designed to meet the installation, operational and/or maintenance needs of companies or the systems they use and operate. Training requirements are used to define the training programs.

Training programs are learning sessions that are structured to increase skill sets of employees. Learning sessions may be performed via instructor led training, online training or independent self paced courses. Training programs may include the assessment or adaptation of course assignments to adjust for the current skill sets of the students.

Acceptance

System acceptance is the formal turn-over of the new network to the customer. As such, it requires a formal set of testing of performance, as well as the completion of administrative and contractual obligations, and approval before acceptance can happen. The list of items is usually created prior to contract negotiations, and included in the contract. Also, a *punch-list* of detailed items is typically developed during implementation that the purchaser can refer to, to allow a complete and successful acceptance to take place.

Procurement Terms and Conditions

Terms and Conditions of Proposal

The conditions of proposal are the terms for the submission of a document or proposal. Examples of conditions of proposal include how much cost (if any) is authorized for the preparation of a proposal, and how confidential or proprietary information will be handled. Other items listed in this section, usually prepared by the legal and procurement departments, include the following (items are briefly explained following the list):

Acceptance
Audit, testing and inspection
Changes to the SOW
Claims, Disputes
Commencement, scheduling and termination of work, early termination
Compliance with laws and regulations
Confidentiality, privacy, NDA
Covenants
Delays
Delivery
Documentation, Drawings
Force Majeure
Hazardous and toxic waste
Indemnification, patent infringement, intellectual property rights
Inspection
Insurance, Bonds
Labor relations
Limitation of liability
Liquidated damages
Notice and communications
Patent, fees, royalties
Payment and transactions
Performance guarantees

Project Management, coordination
   Responsibilities of parties
   Safety and protection
   Shipments
   Site cleanup
   Storage and operations and site cleanup
   Subcontractors, partners, vendors, agents
   Suspension, Default
   Title, title holder, risk of loss
   Upgrades

Change Orders

Change orders are instructions that indicate alterations or modifications to a specification or plan.

Compliance

Compliance is the process of following rules or operating within established parameters.

Nonresponsive Proposal

Nonresponsive proposals are proposal submissions that have omitted required portions or been structured in a format other than the format that was required in the request for proposal. Nonresponsive proposals may be excluded from consideration for the awarding of projects.

When details (specifications) of key devices or systems are not provided, it is called silence of specifications. RFPs may include a clause defining how the details or specifications will be determined when silence of specifications occurs. This may include best possible device or system or through negotiations.

Performance Guarantees

Performance guarantees are commitments from providers of products or services on the performance levels that will be achieved. Performance guarantees usually define what will happen if a performance level is not achieved such as financial penalties. System performance levels may be defined for different types of operational modes, including normal and loaded (stressed) conditions. To ensure that financial penalties will be paid in the event of inability to achieve the committed performance levels, a supplier may be required to provide (pay for) a performance bond.

Proprietary Information

Proprietary information is data or materials that are owned or controlled by another person or company that may be harmful or perceived to be harmful if obtained by unauthorized recipients. Proprietary information may include customer lists, business processes, operational capabilities or technologies. An RFP may specify how proprietary information may be received and controlled by the RFP issuer.

Warranty and Guarantees

Warrantees and guarantees for RFPs include typical equipment return and repair, or, for systems and installations, indemnification and performance bond. Indemnification is the providing of protection against damage, loss or injury. A performance bond is a financial instrument that is used by a company that is purchasing products or services to ensure compensation is paid in the event a product, system or service does not achieve the guaranteed performance objectives.

Regulatory Compliance

Regulatory compliance is the process of following governmental rules for the development or operation of products, services or systems.

Liability

The liability section deals with who is responsible for any damage that might occur. Payment of damages may result from such situations.

Right to Reject

Right to reject is the terms that are associated with the rejection of a proposal. The right to reject may be unconditional (for any reason) or it may contain specific criteria such as nonresponsive or incomplete sections, experience or product support capabilities.

Pricing and Financing Options

Pricing Options

Pricing options include equipment, service, support, and training prices and options. Equipment pricing options should include key types of equipment and the cost of purchasing additional equipment after the RFP requirements are complete. Service pricing options should be included which define communication lines or hosting service fees. Support pricing should define technical support and training courses. Pricing options should also define discounts and penalties, along with their terms.

Pricing options are a set of price plans or rates that are based on qualification or selection criteria. An example of pricing options is the difference in prices that are charged for wholesale and retail customers. The RFP should outline the financial objectives and needs of the company relative to any pricing or financial options response.

Elements in the Business Case

RFP issuers must look to the overall business case to determine what elements to include and exclude from the pricing options required from the respondent. Elements of capital cost for a WiMAX network include CPE investment, installation and commissioning CPE, radio communications and data network communications planning and engineering, base station and other infrastructure elements, tower or rooftop acquisition cost, IP switching and routing costs, backhaul microwave transmission costs, network management and NOC costs, and customer care and billing, if appropriate. Also included are initial customer acquisition costs (advertising) and general and administrative costs. Elements of operating costs include leases for tower or rooftop base station antenna placement, operations and maintenance of CPE, network elements, NOC, leased transmission circuits for backhaul, facilities and office space, ongoing advertising, and facilities management and maintenance.

Equipment Pricing

RFPs for WiMAX Forum typically contain spreadsheets that require completion by the respondent, which enables the issuer to understand the elements of a system solution. This may include capital costs for equipment, equipment rollout by year, equipment by area, or for the base network, plus any options.

Support Services

The financial response spreadsheet may include sections for operating costs, recurring costs, license fees, support fees, service level agreement fees.

Financing Terms

Financing terms are a set of payment options along with the terms (such as payment intervals and interest rates) associated with each payment option. An RFP may include financing requirements that a vendor must provide if the buyer selects their products or services.

Submission of Financial Information

Cost information is typically submitted in electronic spreadsheet format. The format of response is typically provided by the issuer and the respondent must use the template exactly as provided to ensure proper comparison among vendors. The sheet should contain columns for periods (i.e. years) and rows for the cost elements (i.e. capital, other first costs, as well as recurring costs, such as operating and maintenance costs including license fees and services fees). Options can be broken out separately.

Afterword

Creating WiMAX RFPs is an important part of the planning, design, and deployment of a WiMAX network when organizations desire competition for cost, schedule and performance in broadband wireless solutions. Since this book is focused on how to get vendors to expose the best that their products can offer, and not just on any specific current capability of WiMAX, this book should last you for many upgrades and generations of WiMAX and broadband wireless technology to come!

If you wish to dig deeper into WiMAX and broadband wireless network design and design optimization, visit the author's web site for additional books and references that will allow you to do so.

Whenever you need a new or improved WiMAX system or network, and you want to gauge the capabilities of the WiMAX industry as a whole, take this book, contact a good consultant, and start Creating WiMAX RFPs!

Good luck!

Tony Klinkert
Plano (Dallas), Texas
USA

References:

[1] Session Initiation Protocol, http://www.sipforum.org/
[2]. Wi-MAX Technical Information, www.WiMAXforum.org/tech, 2 April, 2004.
[3] IEEE, http://www.ieee.org
[4] WiMAX Forum web site, http://www.WiMAXforum.org/about/
[5] IEEE 802.16 Standards, http://standards.ieee.org/getieee802/802.16.html
[6] See for example: http://www.procuri.com/

Appendix 1

Acronyms

3DES-Triple Data Encryption Standard
AAA-Authentication, Authorization, Accounting
AAS-Adaptive Antenna System
ABR-Available Bit Rate
ACD-Automatic Call Delivery
ACK-Acknowledgment
AES-Advanced Encryption Standard
AMC-Adaptive Modulation and Coding
AP-Access Point
ARQ-Automatic Retransmission Request
ATA-Analog Telephone Adapter
AUD-Authorized Dealer
BE-Best Effort Service
BER-Bit Error Rate
BOM-Bill Of Materials
BPSK-Binary Phase-Shift Keying
BRAN-Broadband Radio Access Networks
BRS-Broadband Radio Service
BS-Base Station
BW-Bandwidth
BWA-Broadband Wireless Access
C/I-Carrier To Interference Signal Ratio
CapEx-Capital Expenditure
CAS-Conditional Access System
CBR-Constant Bit Rate
CDMA-Code Division Multiple Access
Centrex-Central Exchange
CINR-Carrier to interference-plus-noise ratio
COS-Class Of Service
CPE-Customer Premises Equipment
CSMA/CA-Carrier Sense Multiple Access/Collision Avoidance
CSR-Customer Service Representative
CTI-Computer Telephony Integration
DAMA-Demand Assigned Multiple Access
DAS-Distributed Antenna System
DES-Data Encryption Standard
Design-Build-Design Build
DFS-Dynamic Frequency Selection
DHCP-Dynamic Host Configuration Protocol
DMU-Decision Making Unit
DOCSIS+-Data Over Cable Service Interface Specification Plus
DSL-Digital Subscriber Line
DVB-Digital Video Broadcast
EAP-Extensible Authentication Protocol
E-commerce or ECommerce-Electronic Commerce
EE-End To End
EFI-Engineer, Furnish, and Install
EIRP-Effective Isotropic Radiated Power
EPC-Engineer, Procure, and Construct
ETSI-European Telecommunications Standards Institute
FDD-Frequency Division Duplex
FDM-Frequency Division Multiplexing
FDMA-Frequency Division Multiple Access
FEC-Forward Error Correction

FFT-Fast Fourier Transform
FHSS-802.11 Frequency Hopping Spread Spectrum
FOC-Full Operational Capability
FTP-File Transfer Protocol
GPS-Global Positioning System
GSM-Global System For Mobile Communications
HD-High Definition
Hotspots-Hot Spots
Hotspots-Wireless Hot Spots
HSDPA-High Speed Downlink Packet Access
HTTP-Hypertext Transfer Protocol
IAD-Integrated Access Device
ICS-Implementation Conformance Statement
IEEE-Institute Of Electrical And Electronics Engineers
IOC-Initial Operational Capability
IP-Internet Protocol
IP STB-Internet Protocol Set Top Box
IT-Information Technology
IT Integration-Information Technology Integration
IVR-Interactive Voice Response
KEK-Key Encryption Key
LAN-Local Area Network
LMDS-Local Multichannel Distribution Service
LOS-Line Of Sight
LSB-Least Significant Bit
MAC-Medium Access Control
MAN-Metropolitan Area Network
MDU-Multiple Dwelling Unit
MIB-Management Information Base
MIMO-Multiple In Multiple Out
MIS-Management Information System
MPEG-Motion Picture Experts Group
MS-Mobile Station
MTU-Multi Tenant Unit
NDA-Non-Disclosure Agreement

Net-Internet
NI-Network Interface
NIC-Network Interface Card
NLOS-Non-Line of Sight
NOC-National Operations Center
OEM-Original Equipment Manufacturer
OFDM-Optical Frequency Division Multiplexing
OFDM-Orthogonal Frequency Division Multiplexing
OFDMA-Orthogonal Frequency Division Multiple Access
OSI-Open Systems Interconnection
OSS-Operations Support System
OTS-Off the Shelf
PAN-Personal Area Network
PDN-Premises Distribution Network
PDN-Public Data Network
PDU-Protocol Data Unit
PER-Packet Error Rate
PHY-Physical Layer
PICS-Protocol Implementation Conformance Statement
PMP-Point to Multipoint
PO-Purchase Order
PSTN-Public Switched Telephone Network
PTP-Point to Point
PUC-Public Utilities Commission
QAM-Quadrature Amplitude Modulation
QoS-Quality Of Service
QPSK-Quadrature Phase Shift Keying
RAM-Reliability, Availability and Maintainability
RCT-Radio Conformance Tests
RF-Radio Frequency
RFB-Request for Bid
RFI-Request For Information
RFO-Request for Offer
RFP-Request For Proposal
RFQ-Request For Quotation
RFQ-Request for Quote

ROW-Right of Way
RSSI-Received Signal Strength Indicator
SA-Security Association
SAS-Single Award Schedule
SC-Single Carrier
SCADA-System Control and Data Acquisition
SD-Standard Definition
SDMA-Spatial Division Multiple Access
SDR-Service Detail Record
SI-System Integrator
SLA-Service Level Agreement
SNMP-Simple Network Management Protocol
SNR-Signal To Noise Ratio
SOW-Scope of Work
SOW-Statement of Work
SS-Subscriber Station
T-commerce-Television Commerce
TCP-Transmission Control Protocol
TDD-Time Division Duplex
TDM-Time Division Multiplexing
TDMA-Time Division Multiple Access
TOS-Type Of Service
UBR-Unspecified Bit Rate
UDP-User Datagram Protocol
USB-Universal Serial Bus
VLAN-Virtual Local Area Network
VoIP-Voice Over Internet Protocol
vPBX-Virtual PBX
Wi-Fi-Wireless Fidelity
WiMAX-Worldwide Interoperability for Microwave Access
WLAN-Wireless Local Area Network
WMAN-Wireless Metropolitan Area Network
WPAN-Wireless Personal Area Network

Appendix 2

Glossary

IEEE 802.16 – The Institute of Electrical and Electronic Engineers[1] (IEEE) Wireless Metropolitan Area Network (WMAN) standard for broadband wireless technology. This technology is the outdoor version of the popular indoor 802.11[2] Wireless Local Area Network (WLAN) technology, with many enhancements for outdoor operation. The 802.16 technology standard, compared to other wireless technology standards was designed primarily for broadband service, long distance operation, high spectrum efficiency, high capacity, optimal security, and outstanding user choice of service quality from best effort to the highest reliability wireless service commercially available (called quality of service, or QoS).

WiMAX Certification – The WiMAX Forum[3] is an industry association with the primary goal of testing and certifying that equipment built to IEEE 802.16 standards can interoperate among different vendors.

WiMAX Profile – A specific version and set of options of the IEEE 802.16 standard, with a basic set of capabilities, and a specific frequency of operation.

Pre-WiMAX Equipment – Some vendors use this term to indicate that the equipment they are selling or that is in operation has some combination of the same or better capabilities as WiMAX certified equipment or equipment designed according to 802.16 standards, such as in bandwidth or outdoor coverage range. This equipment may not ultimately be WiMAX certified equipment without equipment replacement or software upgrades.

WiMAX Ready Equipment – Some vendors use this term to indicate that the equipment they are selling at the moment may not be WiMAX Certified (due to the fact that the WiMAX Forum lab has yet to create a profile for that particular set of options, capabilities, or frequency of operation), but the vendor is asserting that the equipment can be made to pass the certification at some point in the future.

WiMAX Certified Equipment – Equipment that has passed WiMAX laboratory certification tests, for the specific profile, and meets all the basic interoperability functional capabilities of a specific WiMAX test profile of interest to a buyer.

Access Point (AP)-Access point (AP) is typically a point that is readily accessible to customers for access to a wireless or wired system. It may also be referred to as a radio port or access node.

Advanced Encryption Standard (AES) - Is an encryption standard by a US Government Agency, the National Institute of Standards and Technology (NIST). It is documented in US FIPS PUB 197 (called "FIPS 197"). This encryption is considered "strong encryption", and is practically (but not theoretically) impossible to breach. The National Security Agency (NSA) has approved this encryption technique for top secret information. It is an updated version of the latest Data Encryption Standard (DES), Triple-DES.

Acknowledgment (ACK)-A process or control code that is used to confirm whether a message has been received or a process has been started or completed. An example of an acknowledgement process is the sending of an ASCII 06 character during data transmission between computers to indicate that data has been received without any errors.

Adaptive Antenna System (AAS)- Allows a transmitter to transmit focused radio beams to increase the transmission range, reduce interference and increase signal quality. When an AAS system is used to allow multiple users to communicate with the same transceiver (multiple beams), it is called spatial division multiple access (SDMA).

Adaptive Modulation and Coding (AMC)-The process of dynamically adjusting the modulation type and channel coding of a communication channel based on channel requirements and quality.

Automatic Retransmission Request (ARQ)-An acknowledgment process whereby the sending device can retransmit blocks of data that were received incorrectly by the receiving device.

Available Bit Rate (ABR)-A communications service category that provides the user with a data transmission rate that varies depending upon the availability of the network resources. ABR service may provide the user with feedback as to the changed data transfer rate and may have established minimum and maximum levels of data transmission rates.

Bandwidth (BW)-A term that defines the signals that occupy a portion of a frequency spectrum, particularly a radio system, or the data transmission rate. Analysis or measurement of the signals or signal waveforms of such a system will show that most (or substantially all) of the power contained in that signal can be found in a designated portion of the frequency spectrum. The difference between the highest and lowest frequency describing that portion of the spectrum is the bandwidth of the signal. Frequency (radio spectrum) bandwidth is measured in units of hertz or cycles per second and data transmission bandwidth is measured in bits per second.

Base Station (BS)-The radio part of a mobile radio transmission site (cell site), a single base station usually contains several radio transmitters, receivers, control sections and power supplies. Base stations are sometimes called a land station or cell site.

Best Effort Service (BE)-A level of service in a communications system that doesn't have a guaranteed level of quality of service (QoS).

Binary Phase-Shift Keying (BPSK)-A modulation process that converts binary bits into phase shifts of the radio carrier without substantially changing the frequency of the carrier waveform. The phase of a carrier is the relative time of the peaks and valleys of the sine wave relative to the time of an unmodulated "clock" sine wave of the same frequency. BPSK uses only two phase angles, corresponding to a phase shift of zero or a half cycle (that is, zero or 180 degrees of angle).

Bit Error Rate (BER)-BER is calculated by dividing the number of bits received in error by the total number of bits transmitted. It is generally used to denote the quality of a digital transmission channel.

Broadband Radio Access Networks (BRAN)-A portion of a communication network that allows individual subscribers or devices to connect to the network via radio signals and obtain broadband services (greater than 1 Mbps).

Broadband Radio Service (BRS)-The providing of high speed data transmission services through the use of microwave frequencies around 2.5 GHz (formerly called MMDS).

Broadband Wireless Access (BWA)-A term that is commonly associated with wireless high-speed data transfer connections. When applied to consumer access networks, broadband often refers to data transmission rates of 1 Mbps or higher.

Carrier Sense Multiple Access/Collision Avoidance (CSMA/CA)-A network access method in which each device signals its intent to transmit data before it actually does so. This keeps other devices from signaling information, thus preventing collisions between the signals of two or more devices. CSMA/CA is used on networks, which operates in the unlicensed 2.4 GHz ISM band.

Carrier to Interference Signal Ratio (C/I)- The ratio of desired signal to the amount of interference signals from all unwanted interfering signals. The C/I ratio is commonly expressed in dB. Different types of systems can tolerate different levels of interference dependent on the modulation type and error protection system

Carrier to interference-plus-noise ratio (CINR)- The ratio of the desired signal to the noise floor in the environment from natural and man-made noise sources plus the amount of interference from all unwanted interfering signals at the location. The CINR is commonly expressed in dB. Different types of systems can tolerate different levels of CINR dependent on the modulation type and error protection system

Class Of Service (COS)-(1-multimedia) The communication parameters that are assigned or associated with a particular application or communication session. The class of service usually requires a specific quality of service (QoS) level. (2-telecommunications) Categories of services that are provided by tariffs charged to customers for the particular services they select. Examples of class services include flat rate, coin, toll free (800) service, and PBX. (3-SS7) Services provided by the Signaling Connection Control Pant (SCCP) to its users.

Code Division Multiple Access (CDMA)-The sharing of a radio channel by multiple users by adding a unique code for each data signal that is being sent to and from each of the radio transceivers. These codes are used to spread the data signal to a bandwidth much wider than is necessary to transmit the data signal achieving many benefits.

Constant Bit Rate (CBR)-A class of telecommunications service that provides an end-user with constant bit data transfer rate. CBR service is often used when real-time data transfer rate is required such as for voice service. For example constant bit rate service is useful for high qualifiying audio transmission.

Customer Premises Equipment (CPE)-All telecommunications terminal equipment located on the customer's premises, including telephone sets, private branch exchanges (PBXs), data terminals, and customer-owned coin-operated telephones.

Data Encryption Standard (DES)-An encryption algorithm that is available in the public domain and was accepted as a federal standard in 1976. It encrypts information in 16 stages of substitutions, transpositions and nonlinear mathematical operations.

Data Over Cable Service Interface Specification Plus (DOCSIS+)-A version of Data Over Cable Service Interface Specification (DOCSIS) that is used for wireless broadband.

Data Encryption Standard (DES) - Is an encryption method (known as a cipher) used since the mid-1970's in commercial and government systems. It is currently considered insecure due to a short key length, which makes the cipher vulnerable to compromise. A newer version, Triple DES, is considered practically (although not theoretically) secure. A newer cipher, AES, has been developed and is the recommended cipher for the most secure communications.

Data Throughput - The amount of data information that can be transferred through a communication channel or transfer through a point on a communication system. Gross throughput includes overhead (managed) information. Net throughput ("good put" or "payload") is only the user traffic.

Demand Assigned Multiple Access (DAMA)-A process that allows the sharing of the capacity of a communication channel or groups of communication channels through the assignment of unused channels as the demand for capacity increases.

Digital Subscriber Line (DSL)-The transmission of digital information, usually on a copper wire pair. Although the transmitted information is in digital form, the transmission medium is usually an analog carrier signal (or the combination of many analog carrier signals) that is modulated by the digital information signal.

Digital Video Broadcast (DVB)-The sending of television signals over digital transmission channels. DVB transmission can occur over different types of systems including wimax broadcast radio, satellite systems, cable television systems and mobile communications. DVB industry standards are published by the joint technical committee (JTC) of the European Telecommunications

Standards Institute (ETSI). These standards can be obtained at www.ETSI.org

Dynamic Frequency Selection (DFS)-A process that allows devices or users to request, select or change an operating frequency at various times.

Dynamic Host Configuration Protocol (DHCP)-A process that dynamically assigns an Internet Protocol (IP) address from a server to clients on an as needed basis. The IP addresses are owned or controlled by the server and are stored in a pool of available addresses. When the DHCP server senses that a client needs an IP address (e.g. when a computer boots up in a network), it assigns one of the IP addresses available in the pool.

European Telecommunications Standards Institute (ETSI)– A European standardization organization for the telecommunications industry. ETSI was the body that standardized the GSM cellular standard as well as the earlier TETRA mobile radio system. It is an independent, not for profit, organization based in Sophia Antipolis, France.

Effective Isotropic Radiated Power (EIRP)-The relative power to an antenna compared to the RF signal gain in a given direction of an isotropic antenna.

Extensible Authentication Protocol (EAP)-A standard protocol developed by the IETF that allows for authentication on PPP and on data networks such as Ethernet and WiMAX.

Fast Fourier Transform (FFT)-An alternative implementation of the Discrete Fourier Transform that eliminates significant computational redundancy. Fast Fourier Transform converts signals into their frequency components.

File Transfer Protocol (FTP)-A protocol that is used to manage the transfer of data files between computers and networks. Because FTP is a standard protocol, it permits the transfer of any type of data file between different types of computers or networks.

Forward Error Correction (FEC)-A mathematical algorithm that is used to produce extra bits of data that are sent to each packet. The sending of FEC bits allows for the correction of information bits that were lost or changed due to noise or other physical effects. The extra bits increase the data rate required for a given transmission typically by 5% to 50% depending on the level of correction required so they are only used where error rates are high and retransmission is uneconomical.

Frequency Hopping Spread Spectrum (FHSS) - FHSS uses a radio transmission process where a message or voice communication is sent on a radio channel that regularly changes frequency (hops) according to a predetermined code. The receiver of the message or voice information must also receive on the same frequencies using the same frequency hopping sequence.

Institute of Electrical and Electronics Engineers (IEEE)- A standards body and professional association that is focused on electronics and communications technologies. It is a large organization with over 300,000 members in over 140 countries. Standards it has successfully developed include 802.3 ("Ethernet"), 802.11 (Wireless LAN or "WiFi"), and now 802.16 ("WiMAX").

Frequency Division Duplex (FDD)-The process of allowing the transmission of information in both directions (not necessarily at the same time) via separate bands (frequency division). When using FDD, each device transmits on one frequency and listens on a different frequency.

Frequency Division Multiple Access (FDMA)-The process of allowing mobile radios to share radio frequency allocation by dividing up that allocation into separate radio channels where each radio device can communicate on a single radio channel during communication.

Frequency Division Multiplexing (FDM)-The multiplexing of two or more signals into one output by assigning each signal its own bandwidth within a broad range of frequencies. Frequency division multiplexing is used to divide a frequency bandwidth into several smaller bandwidth frequency channels. Each of these smaller channels is used for one communications channel.

FM radio, broadcast television and some cellular telephone systems use frequency division multiplexing.

Geographic Spectrum Efficency- see Spectrum Efficency

Global Positioning System (GPS)-A network of 24 Navstar satellites that are orbiting the Earth at 11,000 feet above the surface that provide signals allowing the calculation of position information. Each GPS satellite transmits two frequencies; 1575.42 MHz (the L1 carrier) and 1227.6 MHz (the L2 carrier). A GPS receiver compares the signals from multiple GPS satellites (4 satellite signals are usually used) to calculate the geographic position. In March of 1996, the military requirements for limited accuracy transmission were lifted. This allows the GPS system to provide very precise vehicle or device locations.

Global System For Mobile Communications (GSM)-A wide area wireless communications system that uses digital radio transmission to provide voice, data, and multimedia communication services. A GSM system coordinates the communication between mobile telephones (mobile stations), base stations (cell sites), and switching systems. Each GSM radio channel is 200 kHz wide channels that are further divided into frames that hold 8 time slots. GSM was originally named Groupe Speciale Mobile. The GSM system includes mobile telephones (mobile stations), radio towers (base stations), and interconnecting switching systems.

High Speed Downlink Packet Access (HSDPA)-A mobile packet data service that uses the wideband code division multiple access (WCDMA) system to provide downlink data transmission rates of 8-10 Mbps.

Hypertext Transfer Protocol (HTTP)-A protocol that is used to transmit hypertext documents through the Internet. It controls and manages communications between a Web browser and a Web server.

Indoor Base Stations-Devices or assemblies that enable systems or system connections to receive and convert radio or communication signals that are enclosed and/or have connections that allow them to be used in indoor environments.

Indoor Subscriber Stations-Devices or assemblies that enable users to receive and convert radio or communication signals that are enclosed and/or have connections that allow them to be used in indoor environments.

Internet (Net)-A public data network that interconnects private and government computers together. The Internet transfers data from point to point by packets that use Internet protocol (IP). Each transmitted packet in the Internet finds its way through the network switching through nodes (computers). Each node in the Internet forwards received packets to another location (another node) that is closer to its destination. Each node contains routing tables that provide packet forwarding information. The Internet evolved from ARPANET and was designed to allow continuous data communication in the event that some parts of the network were disabled.

Internet Protocol (IP)-A low-level network protocol that is used for the addressing and routing of packets through data networks. IP is the common language of the Internet. The IP protocol only has routing information and no data confirmation rules. To ensure reliable data transfer using IP protocols, higher level protocols such as TCP are used. IP protocol is specified in RFC-791. This protocol defines the packet datagram that holds packet delivery addressing, type of service specification, dividing and re-assembly of long data files and data security. IP protocol structure is usually combined with high-level transmission control protocols such as transaction control protocol (TCP/IP) or user datagram protocol (UDP/IP).

Key Encryption Key (KEK)-A code or value that is used to encrypt another key.

Least Significant Bit (LSB)-In a digital word, the bit that defines the smallest increment of resolution and is usually the last bit in the word sequence.

Line Of Sight (LOS)-A direct path in a wireless communication system that does not have any significant obstructions. LOS systems can use optical or radio signals for transmission.

Local Area Network (LAN)-Private data communication networks that use high-speed digital communications channels for the interconnection of computers and related equipment in a limited geographic area. LANs can use fiber optic, coaxial, twisted-pair cables, or radio transceivers to transmit and receive data signals. LANs are networks of computers, normally personal computers, connected together, in close proximity (office setting), in order to share information and resources. The two predominant LAN architectures are token ring and Ethernet. Other LAN technologies are ArcNet, AppleTalk, and fiber distributed data interface (FDDI).

Local Multichannel Distribution Service (LMDS)-A wireless broadband distribution system that operates in the 28 GHz to 31 GHz frequency band. In the United States, LMDS entered into the FCC auction process in 1997. LMDS uses approximately 1.3 GHz wide spectrum band. This provides a typical data rate for each LMDS channel of 1 Gbps. Because of the extremely high frequencies used, the transmitter must be located within only a few miles of the receiver. The benefit of short distance is that LMDS signals from one cell will not interfere with other cells placed 10 or more miles apart. This allows the radio bandwidth to be reused (frequency reused) in a cellular like fashion.

Management Information Base (MIB)-A collection of definitions, which define the properties of the managed object within the device to be managed. Every managed device keeps a database of values for each of the definitions written in the MIB. MIBs are used in conjunction with the simple network management protocol (SNMP) as well as Remote Monitoring Specification (RMON) to manage networks. MIBs (referred to now as MIB-i) were originally defined in RFC1066.

Medium Access Control (MAC)- The processes used by communication devices to gain access to a shared communications medium or channel. Examples of MAC systems CSMA/CD and Token Passing. A MAC protocol is used to control access to a shared communications media (transmission medium) which attaches multiple devices. MAC is part of the OSI model Data-Link Layer. Each networking technology, for example Ethernet, Token Ring or FDDI, have drastically different protocols which are used by devices to gain access to the network, while still providing an interface that upper layer protocols, such as TCP/IP may use without regard for the details of the technology. In short, the MAC provides an abstract service layer that allows network layer protocols to be indifferent to the underlying details of how network transmission and reception operate.

Metropolitan Area Network (MAN)-A data communications network or interconnected groups of data networks that have the geographic boundaries of a metropolitan area. The network is totally or partially segregated from other networks, and typically links local area networks (LANs) together.

Mobile Station (MS)-A mobile radio telephone operating within a wireless system (typically cellular or PCS). This includes hand held units as well as transceivers installed in vehicles.

Multichannel Multipoint Distribution Service (MMDS)- The providing of television services through the use of 2.5 GHz microwave frequencies. MMDS is commonly called "wireless cable." This service has been updated and renamed Broadband Radio Service (BRS).

Multiple In Multiple Out (MIMO)-The combining or use of two or more radio or telecom transport channels for a communication channel. The ability to use and combine alternate transport links provides for higher data transmission rates (inverse multiplexing) and increased reliability (interference control). See also AAS

Non-Line of Sight (NLOS)-A wireless communication system that does not require a direct path (can have significant obstructions) between the transmitter and receiver. NLOS systems can use optical or radio signals for transmission. Current techniques for delivery of NLOS are spread spectrum techniques, mesh topologies, and/or OFDM modulation schemes.

Open Systems Interconnection (OSI)-Standard layer model that was developed by the international standards organization (ISO) and the CCITT. The OSI model helps to standardize the inter-connection of computers and data terminals to their applications, regardless of their type or manufacturer. The protocols specify seven layers: physical, link, network, transport, session, presentation, and application. Each layer performs specific functions for data exchange and is independent of the other layers.

Operational Support System (OSS)-Combinations of equipment and software that are used to allow a network operator to perform the administrative portions of the business. These functions include customer care, inventory management and billing. Originally, OSS referred to the systems that only supported the operation of the network. The recent definition includes all systems required to support the communications company including network systems, billing, customer care, etc.

Optical Frequency Division Multiplexing (OFDM)-A process of transmitting several high speed communication channels through a single fiber with the use of separate wavelengths (optical frequencies) for each communication channel. OFDM is now commonly called wave division multiplexing (WDM). However, WDM usually refers to optical channels that have very small spacing between them and OFDM refers to multiple optical channels that can have any amount of wavelength spacing between them.

Orthogonal Frequency Division Multiple Access (OFDMA)-A process of dividing a radio carrier channel into several independent subcarrier channels that are shared between simultaneous users of the radio carrier. When a mobile radio communicates with an OFDMA system, it is assigned a specific subcarrier channel or group of subcarrier channels within the radio carrier. By allowing several users the use of different subcarrier channels, OFDMA systems increase their ability to serve multiple users and the OFDMA system may dynamically allocate varying amounts of transmission bandwidth based on how many subcarrier channels have been assigned to each user.

Orthogonal Frequency Division Multiplexing (OFDM)-A process of simultaneously transmitting several communication channels through the use of different frequencies, whereby each communication channel is independently managed and optimized.

Outdoor Base Stations-Devices or assemblies that enable systems or system connections to receive and convert radio or communication signals that are enclosed and/or have connections that allow them to be used in outdoor environments.

Outdoor Subscriber Stations- Outdoor subscriber stations are devices or assemblies that enable users or devices to receive and convert radio or communication signals that are enclosed and/or have connections that allow them to be used in outdoor environments.

Packet Error Rate (PER)-Packet error rate is calculated by dividing the number of packets received in error by the total number of bits transmitted. It is generally used to denote the quality of a digital transmission channel.

Personal Area Network (PAN)- Short-range communication networks that typically have a range of 5 to 25 feet. A PAN is used to connect terminal equipment (mice and keyboards) to computing and data network equipment.

Physical Layer (PHY)-Performs the conversion of data to a physical medium (such as copper, radio, or optical) transmission and coordinates the transmission and reception of these physical signals. The physical layer receives data for transmission from an upper layer, such as the Open System Interconnection (OSI) Data Link layer, and converts it into physical format suitable for transmission through a network (such as bursts, slots, frames, and superframes). An upper layer provides the physical layer with the necessary data and control (e.g. maximum packet size) to allow conversion to a format suitable for transmission on a specific network type and transmission line. The physical layer is layer 1 in the OSI protocol layer model.

Point to Multipoint (PMP)-The transfer of information from one device (or point) to multiple points or devices (multiple receiving points). PMP services are broadcast or multicast services.

Point to Point (PTP)-The transmission of signals from one specific point to another. Point-to-point communication uses addressing to deliver information to a specific receiver of the information. It is possible to implement point-to-point communication through a broadcast network by using device addressing or through a network using network routing.

Protocol Data Unit (PDU)-A package of data (group of data bits) that contains header and data protocol information that is used to communicate with a specific protocol layer in a software stack.

Protocol Implementation Conformance Statement (PICS)- A document that is provided by a company or testing facility that states that the product or system provides and/or supports a specific set of protocols.

Quadrature Amplitude Modulation (QAM)-A combination of amplitude modulation (changing the amplitude or voltage of a sine wave to convey information) and phase modulation. There are several ways to build a QAM modulator. In one process, two modulating signals are derived by special pre-processing from the information bit stream. Two replicas of the carrier frequency sine wave are generated; one is a direct replica and the other is delayed by a quarter of a cycle (90 degrees). Each of the two different derived modulating signals are then used to amplitude modulate one of the two replica carrier sinewaves, respectively. The resultant two modulated signals can be added together. The result is a sine wave having a constant unchanging frequency, but having an amplitude and phase that both vary to convey the information. At the detector or decoder, the original information bit stream can be reconstructed. QAM conveys a higher information bit rate (bits per second) than a BPSK or QPSK signal of the same bandwidth, but is more affected by interference and noise as well.

Quadrature Phase Shift Keying (QPSK)-A type of modulation that uses 4 different phase shifts of a radio carrier signal to represent the digital information signal. These shifts are typically +/- 45 and +/- 135 degrees.

Quality Of Service (QoS)-One or more measurements of the desired performance and priorities of a communications system. QoS measures may include service availability, maximum bit error rate (BER), minimum committed bit rate (CBR) and other measurements that are used to ensure quality communications service.

Radio Conformance Tests (RCT)- A set of procedures that are used to evaluate the operation and performance of the radio part of wireless products or systems.

Radio Frequency (RF)-Those frequencies of the electromagnetic spectrum normally associated with radio wave propagation. RF sometimes is defined as transmission at any frequency upon which coherent electromagnetic energy radiation is possible, usually above 150kHz.

Received Signal Strength Indicator (RSSI)-A signal level indicator, usually on a mobile radio, that regularly displays the approximate level of a received signal. The

RSSI indicator allows a user to determine if the radio signal strength in that area is sufficient to initiate or complete a call.

RF Power Control-A process of adjusting the power level of a mobile radio as it moves closer and further from a transmitter. RF power control is typically accomplished by sensing the received signal strength level and relaying power control messages from a transmitter to the mobile device with commands that are used to increase or decrease the mobile device's output power level.

System Control and Data Acquisition (SCADA) - A type of network and signaling that collects data from sensors in remote locations and sends them to central or distributed computing systems for management and control. These systems are typically used by utility companies for the control of devices on the electric, gas, water, or other service grid.

Security Association (SA)-The establishment of a relationship between two network elements that ensures that traffic passing through the interface is cryptographically secure (typically, through encryption.)

Security Sublayer-The functional process within a communication device or system that performs access controls, identity validation and/or encryption of data.

Service Level Agreement (SLA)- An agreement between a customer and a service provider that defines the services provided by the carrier and the performance requirements of the customer. The SLA usually includes fees and discounts for the services based on the actual performance level received by the customer.

Signal To Noise Ratio (SNR)-A comparison of the information-carrying signal power to the noise power in a system. In a phone call, the signal would be your voice, while the noise could be static, clicking from relays, or any other sound that is not voice. Analogous effects occur in optical networks. SNR is often used as one measure of signal quality.

Simple Network Management Protocol (SNMP)-A standard protocol used to communicate management information between the network management stations (NMS) and the agents (ex. routers, switches, network devices) in the network elements. By conforming to this protocol, equipment assemblies that are produced by different manufacturers can be managed by a single program. SNMP protocol is widely used via Internet protocol (IP) and operates over UDP well-known ports of 161 and 162. SNMP was originally defined in RFC1098 and is now obsolete and has been updated by RFC1157.

Single Carrier (SC)-The use of a single carrier wave that is modified to carry (transport) all of the information. See Multicarrier OFDM

Spatial Division Multiple Access (SDMA)-A system access technology that allows a single transmitter location to provide multiple communication channels by dividing the radio coverage into focused radio beams that reuse the same frequency. To allow simultaneous access, each mobile radio is assigned to a focused radio beam. These radio beams may dynamically change with the location of the mobile radio. SDMA technology has been successfully used in satellite communications for several years.

Spectral Efficiency-A measurement characterizing a particular modulation and coding method that describes how much information can be transferred in a given symbole of information. This is often given as bits per second per Hertz. Modulation and coding methods that have high spectral efficiency also typically are very sensitive to small amounts of noise and interference, and often have low geographic spectral efficiency. (See also Geographic Spectral Efficiency)

Subscriber Station (SS)-Stations that receive and convert radio or other types of communication signals into user information.

Time Division Duplex (TDD)-A process of allowing two way communications between two devices by time sharing. When using TDD, one device transmits, device 1, while the other device listens, device 2, for a short period of time. After the transmission is complete, the devices reverse their roles so device 1 becomes a receiver and device 2 becomes a transmitter. The process continually repeats itself so data appears to flow in both directions simultaneously.

Time Division Multiple Access (TDMA)-A process of sharing a single radio channel by dividing the channel into time slots that are shared between simultaneous users of the radio channel. When a mobile radio communicates with a TDMA system, it is assigned a specific time position on the radio channel. By allowing several users to use different time positions (time slots) on a single radio channel, TDMA systems increase their ability to serve multiple users with a limited number of radio channels.

Time Division Multiplexing (TDM)-A method used to send two or more signals over a common transmission path by assigning the path sequentially to each signal, each assignment at a discrete time interval. All channels of a time-division multiplex system use the same portion of the transmission links' frequency spectrum - but not at the same time. Each channel is sampled in a regular sequence by a multiplexer. See also TDD

Transmission Control Protocol (TCP)-A session layer protocol that coordinates the transmission, reception, and retransmission of packets in a data network to ensure reliable (confirmed) communication. The TCP protocol coordinates the division of data information into packets, adds sequence and flow control information to the packets, and coordinates the confirmation and retransmission of packets that are lost during a communication session. TCP utilizes Internet Protocol (IP) as the network layer protocol.

Triple Data Encryption Standard (3DES)-A variation of the data encryption standard that adds complexity to the encryption process by increasing the difficulty to break the encryption process.

Type Of Service (TOS)-The field within the IP datagram header that indicates the type of packet, so the system can vary the type of service (typically the priority of routing) that is performed on the IP data packet. The TOS field as specified in RFC 760 and RFC 2475 defines an application of TOS as used in DiffServ networks.

Unspecified Bit Rate (UBR)-A category of telecommunications service that provides an unspecified data transmission rate of service to end user applications. Applications that use UBR services do not require real-time interactivity nor do they require a minimum data transfer rate. UBR applications may not require the pre-establishment of connections. An example of a UBR application is Internet web browsing.

User Datagram Protocol (UDP)-A high-level communication protocol that coordinates the one-way transmission of data in a packet data network. The UDP protocol coordinates the division of files or blocks of data information into packets and adds sequence information to the packets that are transmitted during a communication session using Internet protocol (IP) addressing. This allows the receiving end to receive and re-sequence the packets to recreate the original data file or block of data that was transmitted. UDP adds a small amount of overhead (control data) to each packet relative to other high-level protocols such as TCP. However, UDP does not provide any guarantees to data delivery through the network. UDP protocol is defined in request for comments 768 (RFC 768).

Virtual Local Area Network (VLAN)-A data communications network that interconnects computers and related equipment in a limited geographic area where multiple logically setup networks (virtual networks) can be setup and managed through the underlying communication network. VLAN connections are setup to allow data to safely and privately pass over other types of data networks (such as the Internet).

Voice Over Internet Protocol (VoIP)-A process of sending voice telephone signals over the Internet or other data network. If the telephone signal is in analog form (voice or fax) the signal is first converted to a digital form. Packet routing information is then added to the digital voice signal so it can be routed through the Internet or data network.

WiMAX Forum-A non-profit industry group that was created to assist in the development, certification and promotion of IEEE 802.16 and ETSI HiperMAN standards.

Wireless Fidelity (Wi-Fi)-Another name for the 802.11 wireless LAN system.

Wireless Local Area Network (WLAN)-A network that allows computers and workstations to communicate with each other using radio propagation as the transmission medium. The wireless LAN can be connected to an existing wired LAN as an extension, or can form the basis of a new network. While adaptable to both indoor and outdoor environments, wireless LANs are especially suited to indoor locations such as office buildings, manufacturing floors, hospitals and universities.

Wireless Personal Area Network (WPAN)-Temporary (ad-hoc) short-range wireless communication systems that typically connect personal accessories such as headsets, keyboards, and portable devices to communications equipment and networks.

Worldwide Interoperability for Microwave Access (WiMAX)-A wireless communication system that allows computers and workstations to connect to high-speed data networks (such as the Internet) using radio waves as the transmission medium with data transmission rates that can exceed 155 Mbps for each radio channel. The WiMAX system is defined in a group of IEEE 802.16 industry standards and its various revisions are used for particular forms of fixed and mobile broadband wireless access.

References:

[1] See here: www.ieee.org, And here: www.ieee802.org/16/

[2] See: http://www.ieee802.org/11/

[3] The web address for the forum is: www.wimaxforum.org with objectives here: http://www.wimaxforum.org/about

Appendix 3

Wimax Resources

WiMAX RFP support
www.CreatingWiMAXrfps.com – Reader's forum, support for creating RFPs, and recommended books for further learning.

WiMAX
WiMAX Forum: http://www.wimaxforum.org/home
WiMAX Basic Information: http://en.wikipedia.org/wiki/WiMAX
IEEE 802.16 - http://wirelessman.org/
Wireless On Line - http://www.wirelessdesignonline.com
RF Design Guide: http://www.cdt21.com/resources/default.asp

WiMAX Support
WCA - http://www.wcai.com/
WiMAX.Com - http://www.wimax.com/
WiMAX Industry: http://www.wimax-industry.com/
WiMAX 360 - http://www.wimax360.com/
WiMAX Trends - http://www.wimaxtrends.com/
WiMAX Vision - http://www.wimax-vision.com/
WiMAX Networking News - http://wimaxnetnews.com/
WiMAX World - http://www.wimaxworld.com/2008/
WiMAX Day - http://www.wimaxday.net/

WiMAX Vendors
Adaptix (http://www.adaptix.com/)
Airspan (http://www.airspan.com/)
Alcatel – Lucent (http://www1.alcatel-lucent.com/serviceproviders/wimax/)
Alvarion (http://www.alvarion.com/)

Aperto Networks (http://www.apertonet.com/)
Axxcelera (http://www.axxcelera.com/)
Belair Networks (http://www.belairnetworks.com/)
Fujitsu http://www.fujitsu.com/us/services/edevices/microelectronics/broadbandwireless/)
Firetide (www.Firetide.com)
Huawei Technologies (http://www.huawei.com/products/wimax.do)
Infinet Wireless (http://www.infinetwireless.com/)
Intracom Telecom (http://www.intracom-telecom.com/)
Kapsch (http://www.kapsch.net/)
Motorola (http://www.motorola.com/)
Navini-Cisco (http://www.navini.com/)
NEC (http://www.nec.com/global/solutions/nsp/WiMAX/)
Nera (http://www.nera.no/en/)
Nortel (http://www.nortel.com/)
Proxim (http://www.proxim.com/)
Redline (http://www.redlinecommunications.com/)
Runcom (http://www.runcom.com/)
Samsung (http://www.samsung.com/global/business/telecomm/)
Selex Communications (http://www.selex-comms.com/en/)
Siemens (http://www.siemens.ie/carrier/topics/wimax/)
Skypilot (http://www.skypilot.com/)
Soma Networks (http://www.somanetworks.com/)
SR Telecom (http://www.srtelecom.com/)
WiNetworks (http://www.winetworks.com/)
ZTE (http://wwwen.zte.com.cn/)

WiMAX Consultants and System Integrators

Anderson Consulting, now Accenture (http://www.accenture.com/)
Adasta (http://www.adestagroup.com/adesta/)
Barik (http://www.barik.es/)
Bechtel (http://www.bechtel.com/)
Bii Group (http://www.biigroup.com/consulting.asp)
Booz Allen Hamilton (http://www.boozallen.com/)

Comba Telecom Systems (http://www.comba-telecom.com/)
IBM (http://www.ibm.com/)
LCC (http://www.lcc.com/)
PA Consulting Group (http://www.paconsulting.com/)
WFI (http://www.wfinet.com/)

Appendix 4

Wimax RFP Template

I. Summary Needs and Purpose Statement
    a. Executive summary of scope and interest for the RFP
    b. Purpose statement on the intent of the RFP process

II. General Procedures
    a. RFP Submission Procedure
    b. RFP Transmittal Process

III. Issuing Company Information
    a. RFP requirements summary
    b. Company background
    c. Existing systems and services
    d. RFP objectives
    e. Scope of work
    f. Instruction to respondents
    g. RFP distribution
    h. Bidders conference
    i. RFPs clarifications
    j. Responses requirements
    k. Implementation schedule
    l. Subscriber forecasts

IV. Respondent Information
   a. Company background
   b. Qualifications
   c. Experience
   d. Respondent's references
   e. Respondent's support capabilities
   f. Supporting vendors and credentials

V. WiMAX Requirements

   a. Business Requirements
   b. Technical Requirements
   c. Coverage Requirements
   d. Capacity Requirements
   e. Frequency Plan/Interference Mitigation Requirements
   f. Broadband Data Access
   g. Internet Access
   h. Voice and Telephony
   i. Video and Television Services
   j. Capacity Requirements
   k. Frequency Plan & Interference
   l. Distribution System
   m. Access Devices
   n. Security
   o. Technology Refresh and Roadmap
   p. Testing
   q. AAA and CAS
   r. Business Support and Billing System
   s. System Administration and Operational Support
   t. Maintenance, Spares and Repair Parts
   u. Customer Care
   v. Disaster Recovery

VI. Implementation Schedule

    a. Initial Operational Capability
    b. Alpha Testing
    c. Field (Beta) Testing
    d. Full Operational Capability
    e. System Cutover
    f. Acceptance Testing
    g. Training requirements
    h. Pilot, Proof of Concept, or Trial

VII. Procurement Terms and Conditions

    a. Terms and Conditions
    b. Change Orders
    c. Compliance
    d. Nonresponsive proposals
    e. Performance guarantees
    f. Proprietary information
    g. Warranty
    h. Liability
    i. Regulatory compliance
    j. Right to reject

VIII. Pricing and Financing Options

    a. Equipment pricing
    b. Support services
    c. Financing Terms

See Also www.CreatingWIMAXrfps.com for an updated and expanded template

Index

802.11 Wireless LAN, 23
802.16, 23-24, 26, 31-32, 105
802.16e, 27
Acceptance Test, 75, 95
Acceptance Testing, 55, 73, 75, 90
Access Devices, 2, 11, 50-52, 56, 90
Advanced Encryption Standard (AES), 32
Advisory Committee, 67
AES Encryption, 32
Alpha Testing, 73-74, 90
Analog Telephone Adapter (ATA), 15
Approval Authority, 1, 69
Approval Process, 1, 80
Audit, 99
Authentication, Authorization, Accounting (AAA), 55, 90, 95
Authorized Agent, 56
Automatic Call Delivery (ACD), 57
Award Date, 72
Backhaul Circuits, 52
Bake Off, 4
Balance Sheet, 82
Bandwidth (BW), 15, 19, 29, 36, 41
Base Station (BS), 10, 16, 19, 22, 25, 41, 44, 47, 49, 51, 103
Battery Backup, 60-61
Beta Testing, 74, 90
Bill of Materials (BOM), 8-9
Billing System, 58-59, 90, 95
Binary Phase-Shift Keying (BPSK), 41
Bit Error Rate (BER), 60
Broadband, 14-17, 19-20, 22-25, 29-30, 35, 66, 89, 105
Broadband Wireless, 16-17, 19-20, 23-25, 30, 35, 105
Cable Modem Service, 25
Canopy, 40
Cellular Data, 19
Central Exchange (Centrex), 35, 49
Certification Profiles, 32
Certifications, 7, 19, 21, 64-65
Change Orders, 90, 100
Clarification Questions, 72, 81
Clarification Requests, 62, 70-71, 81
Clarification Responses, 71
Clarification Updates, 82
Closing Submission Date, 71
Clutter, 40-41
Commissioning, 55, 84, 103
Company Background, 78, 89, 92, 94
Competitive Sourcing, 4
Compliance, 4, 10, 31, 75-76, 90-91, 99-100, 102
Computer Telephony Integration (CTI), 57
Concentration Point, 25
Conditional Access System (CAS), 2, 11, 56, 90, 95
Conditions of Proposal, 99
Continuity of Coverage, 40
Contract Authority, 69
Contract Negotiation Date, 72
Contracting Authority, 70
Contractual, 5, 41, 98
Coverage, 8, 25, 27, 35-37, 39-44, 46-47, 51, 89
Coverage Objective, 39, 46
Coverage Requirements, 36, 39, 43, 89
Criteria Weighting, 84
Criticality, 53, 60
Customer Care, 2, 11, 57-59, 66, 90, 103
Customer Premises Equipment (CPE), 51-52, 103
Customer Self Care, 58
Customer Service Representative (CSR), 57-58
Cutover, 73, 75, 90, 95
Data Transmission Lines, 24
Date of Submission, 70

Decision Criteria, 1, 13
Decision Matrix, 1, 13, 84
Depth of Coverage, 40-41
Design Build (Design-Build), 9
Design-build, 9
Diagnostics, 54-55
Digital Subscriber Line (DSL), 5, 18, 25, 35, 66
Digital Television Services, 16
Digitized Audio, 15
Disaster Recovery, 2, 61, 90, 95
Distributed Antenna System (DAS), 44
Distribution Plant, 52, 66
DSL Service, 5, 25
Due-Diligence, 28
Duplexing Protocols, 32
Eligible Vendor, 12, 85
Emergency Response Systems, 60
End to end, 8
End To End (EE), 8
Engineer, Furnish, and Install (EFI), 9
Engineer, Procure, and Construct (EPC), 9
Equipment Bid, 8
Ethernet, 23, 51
European Telecommunications Standards Institute (ETSI), 23
Evaluation Criteria, 13, 82
Extensible Authentication Protocol (EAP), 27
Extent of Coverage, 40
Fete Accompli, 28
Field Tools, 54
Field Trials, 74
Final Signing, 72
Financial Information, 64, 94, 104
Financial Modeling, 37
Financial Projections, 64
Financial Qualifications, 64, 82
Financial Strength, 84
Financing Options, 6, 91, 102
Financing Requirements, 6, 77, 104

Force Majeure, 99
Franchise, 58
Franchise Fees, 58
Frequency Channels, 27
Frequency Division Duplex (FDD), 32
Full Operational Capability (FOC), 73, 75, 90, 95
Functional Tests, 96
High Definition (HD), 50
HiperMAN, 23
Hot Spots (Hotspots), 27
Hotspots, 27
Hypertext Transfer Protocol (HTTP), 105
IEEE 802.16, 23-24, 31-32, 105
Implementation Conformance Statement (ICS), 65
Implementation Plan, 73
Indemnification, 99, 101
Indoor Coverage, 41, 43, 51
Industry Standards, 20, 95
Initial Operational Capability (IOC), 74, 90, 95
Initial Operational Capability Date, 74
Interactive Voice Response (IVR), 58
Interconnecting Lines, 24
Internet Access, 5, 14-16, 20, 35, 39, 47, 51, 89
Internet Gateway, 47
Internet Protocol (IP), 5, 15-16, 56, 103
Internet Protocol Set Top Box (IP STB), 56
Interoperability, 20-21, 23-24, 31-32
Issuing Authority, 69
Leased Lines, 35
Legacy, 47, 93
License Exempt Spectrum, 28-29
Licensed, 19, 27-30, 33, 44, 82
Licensee, 65
Line Of Sight (LOS), 26, 41
Link Budget, 44-45
Local Area Network (LAN), 14, 23, 51
Metropolitan Area Network (MAN), 24

Index

Middleware, 11
Milestones, 77, 95
Modulation, 22, 41-42
Motion Picture Experts Group (MPEG), 16
Mounting Assets, 44
Multi Tenant Unit (MTU), 51
Multimedia, 14, 29, 36, 49, 51
Multiple Dwelling Unit (MDU), 51
National Operations Center (NOC), 52, 103
Needs Assessment, 1, 66
Network Architecture, 52, 54, 60
Network Interface (NI), 27, 51, 66
Network Interface Card (NIC), 27, 51
Nomadic Service, 27
Non-Competitive Sourcing, 3-4
Non-Disclosure Agreement (NDA), 77, 99
Nonresponsive Proposals, 90, 100
Notification of Award, 1, 87
Obstacles, 41
Operating Information, 64
Operational Requirements, 5, 21
Optimal Solutions, 12
Outdoor Coverage, 43
Outsourcing, 57
Packet Error Rate (PER), 12, 14, 19, 30, 38, 49
Penetration, 16, 40
Performance Bond, 101
Performance Guarantees, 90, 99, 101
Performance Tests, 96
Pico-Cells, 43
Pilot, 4, 87, 90
Point to Multipoint (PMP), 25
Point to Point (PTP), 15, 20, 25, 52
Power Saving Sleep Modes, 27
Premises Distribution Network (PDN), 51
Pre-Proposal Review, 80
Pre-Qualification Application, 12
Pre-Response Conference, 71
Pre-WiMAX, 24

Pricing Options, 102-103
Prime contractors, 9, 79
Private Data Network, 14
Professional Licenses, 65
Professional services, 9
Profitability, 82
Program Management, 4, 9, 55
Programming Sources, 47
Proprietary Information, 59, 90, 99, 101
Prospective Suppliers, 76
Protest of Award, 88
Public Data Network (PDN), 51
Public Data Networks, 14
Public Notice, 69
Public Switched Telephone Network (PSTN), 14, 16, 22, 49, 52
Purchase Order (PO), 62
Purpose Statement, 88, 91
Quadrature Amplitude Modulation (QAM), 41
Qualification, 1, 7, 12, 65, 75-76, 102
Qualification Requirements, 12
Quality Of Service (QoS), 32, 54
Radio Frequency (RF), 25, 46
Radio Link, 22, 41, 44
Radio Signal Propagation, 41
Radio Spectrum, 21, 26
Radio System, 17, 21-22
Redundancy, 60-61
Refresh, 4-5, 90
Regulatory Compliance, 90, 102
Regulatory Fees, 59
Reliability, Availability and Maintainability (RAM), 60
Remote Administration, 57
Request For Information (RFI), 1, 4, 8, 29, 53, 64
60, 62-65, 67-73, 75-82, 84-85, 87-89, 91-96, 100-104
Request For Quotation (RFQ), 1, 4, 7-8, 12, 64

Request for Quote (RFQ), 1, 4, 7-8, 12, 64
Requirement Dates, 70
Requirements Statements, 9
Reseller, 9
RFP Invite, 62, 76-77
RFP Issuance, 76, 79, 92
RFP Issuer, 1, 6, 37, 62, 77, 88, 91, 94, 101
RFP Objectives, 11-12, 89, 93
RFP Release Date, 71
RFP Requirement Dates, 70
RFP Requirements, 48, 62, 79, 89, 92, 102
RFP Respondent, 38
RFP Response Evaluation, 82
RFP Response Procedure, 75
RFP Response Review, 81
RFP Responses, 80, 82, 84, 91
RFP Submission Procedure, 88, 91
RFP Transmittal, 88, 92
Right of Way (ROW), 66
Right to Reject, 90, 102
Rollout, 103
Royalties, 58, 99
Scalability, 53
Scope of Work (SOW), 41, 89, 93, 99
Self-Diagnostics, 54-55
Service Detail Record (SDR), 59
Service Launch, 12
Service Level Agreement (SLA), 103
Service Provisioning, 56
Silence of Specifications, 100
Single Point of Failure, 60
Site Review, 71
Site Survey, 46
Skill Sets, 64, 98
Spectrum, 5, 16-17, 19, 21, 26-30, 32, 34, 36, 44, 51, 82
Standard Definition (SD), 50
State of the Art, 47
Statement of Work (SOW), 41, 93, 99
Stay of Procurement, 88
Stress Tests, 96
Submission Date, 71
Subscriber Station (SS), 16
Subscriber Unit, 22, 39
Support Systems, 9, 47
Supporting Vendors, 59, 65, 67, 89
System Administration, 2, 56-57, 90, 95
System Control and Data Acquisition (SCADA), 35
System Cutover, 73, 75, 90, 95
System Diagnostics, 54-55
System Gain, 44-45
System Integrator (SI), 9, 12, 47, 49-50
System Profiles, 32
System Redundancy, 60
System Reliability, 60-61
System solution, 8, 40, 103
System Solutions, 8-9
Technical Qualifications, 64
Techno-Economic Analysis, 37
Technology Refresh, 5, 90
Technology standard, 23, 31-32
Television Commerce (T-commerce), 11, 47, 59
Terminals, 22, 27, 51
Testing Requirements, 54, 96
The Last Mile, 46
Time Division Duplex (TDD), 32
Time or Frequency Division Duplex, 32
Topologies, 25
Type Of Service (TOS), 39
Universal Serial Bus (USB), 51
Vendor Selection, 9, 12, 86
Virtual PBX (vPBX), 35, 49
Visioning Session, 67
Voice Over Internet Protocol (VoIP), 5, 9, 15, 47, 49
Voice Service, 35
WiFi, 14, 27, 33, 35
WiMAX Certification, 31-34
WiMAX Certified Equipment, 31

Index

WiMAX Forum, 23-24, 31-33, 105
WiMAX Profile, 30, 34
Winning Responder, 13, 85
Wireless Fidelity (Wi-Fi), 14, 27, 33, 35
Wireless Hot Spots (Hotspots), 27
Wireless Metropolitan Area Network (WMAN), 24, 26

Althos Publishing Book List

Product ID	Title	# Pages	ISBN	Price	Copyright
Billing					
BK7781338	Billing Dictionary	644	1932813381	$39.99	2006
BK7781339	Creating RFPs for Billing Systems	94	193281339X	$19.99	2007
BK7781373	Introduction to IPTV Billing	60	193281373X	$14.99	2006
BK7781384	Introduction to Telecom Billing, 2nd Edition	68	1932813845	$19.99	2007
BK7781343	Introduction to Utility Billing	92	1932813438	$19.99	2007
BK7769438	Introduction to Wireless Billing	44	097469438X	$14.99	2004
IP Telephony					
BK7781361	Tehrani's IP Telephony Dictionary, 2nd Edition	628	1932813616	$39.99	2005
BK7781311	Creating RFPs for IP Telephony Communication Systems	86	193281311X	$19.99	2004
BK7780530	Internet Telephone Basics	224	0972805303	$29.99	2003
BK7727877	Introduction to IP Telephony, 2nd Edition	112	0974278777	$19.99	2006
BK7780538	Introduction to SIP IP Telephony Systems	144	0972805389	$14.99	2003
BK7769430	Introduction to SS7 and IP	56	0974694304	$12.99	2004
BK7781309	IP Telephony Basics	324	1932813098	$34.99	2004
BK7780532	Voice over Data Networks for Managers	348	097280532X	$49.99	2003
IP Television					
BK7781334	IPTV Dictionary	652	1932813349	$39.99	2006
BK7781362	Creating RFPs for IP Television Systems	86	1932813624	$19.99	2007
BK7781355	Introduction to Data Multicasting	68	1932813551	$19.99	2006
BK7781340	Introduction to Digital Rights Management (DRM)	84	1932813403	$19.99	2006
BK7781351	Introduction to IP Audio	64	1932813519	$19.99	2006
BK7781335	Introduction to IP Television	104	1932813357	$19.99	2006
BK7781341	Introduction to IP Video	88	1932813411	$19.99	2006
BK7781352	Introduction to Mobile Video	68	1932813527	$19.99	2006
BK7781353	Introduction to MPEG	72	1932813535	$19.99	2006
BK7781342	Introduction to Premises Distribution Networks (PDN)	68	193281342X	$19.99	2006
BK7781357	IP Television Directory	154	1932813578	$89.99	2007
BK7781356	IPTV Basics	308	193281356X	$39.99	2007
BK7781389	IPTV Business Opportunities	232	1932813896	$24.99	2007
Legal and Regulatory					
BK7781378	Not so Patently Obvious	224	1932813780	$39.99	2006
BK7780533	Patent or Perish	220	0972805338	$39.95	2003
BK7769433	Practical Patent Strategies Used by Successful Companies	48	0974694339	$14.99	2003
BK7781332	Strategic Patent Planning for Software Companies	58	1932813322	$14.99	2004
Telecom					
BK7781316	Telecom Dictionary	744	1932813160	$39.99	2006
BK7781313	ATM Basics	156	1932813136	$29.99	2004
BK7781345	Introduction to Digital Subscriber Line (DSL)	72	1932813454	$14.99	2005
BK7727872	Introduction to Private Telephone Systems 2nd Edition	86	0974278726	$14.99	2005
BK7727876	Introduction to Public Switched Telephone 2nd Edition	54	0974278769	$14.99	2005
BK7781302	Introduction to SS7	138	1932813020	$19.99	2004
BK7781315	Introduction to Switching Systems	92	1932813152	$19.99	2007
BK7781314	Introduction to Telecom Signaling	88	1932813144	$19.99	2007
BK7727870	Introduction to Transmission Systems	52	097427870X	$14.99	2004
BK7780537	SS7 Basics, 3rd Edition	276	0972805370	$34.99	2003
BK7780535	Telecom Basics, 3rd Edition	354	0972805354	$29.99	2003
BK7780539	Telecom Systems	384	0972805397	$39.99	2006

For a complete list please visit
www.AlthosBooks.com

Althos Publishing Book List

Product ID	Title	# Pages	ISBN	Price	Copyright
Wireless					
BK7769431	Wireless Dictionary	670	0974694312	$39.99	2005
BK7769434	Introduction to 802.11 Wireless LAN (WLAN)	62	0974694347	$14.99	2004
BK7781374	Introduction to 802.16 WiMax	116	1932813748	$19.99	2006
BK7781307	Introduction to Analog Cellular	84	1932813071	$19.99	2006
BK7769435	Introduction to Bluetooth	60	0974694355	$14.99	2004
BK7781305	Introduction to Code Division Multiple Access (CDMA)	100	1932813055	$14.99	2004
BK7781308	Introduction to EVDO	84	193281308X	$14.99	2004
BK7781306	Introduction to GPRS and EDGE	98	1932813063	$14.99	2004
BK7781370	Introduction to Global Positioning System (GPS)	92	1932813705	$19.99	2007
BK7781304	Introduction to GSM	110	1932813047	$14.99	2004
BK7781391	Introduction to HSPDA	88	1932813918	$19.99	2007
BK7781390	Introduction to IP Multimedia Subsystem (IMS)	116	193281390X	$19.99	2006
BK7769439	Introduction to Mobile Data	62	0974694398	$14.99	2005
BK7769432	Introduction to Mobile Telephone Systems	48	0974694320	$10.99	2003
BK7769437	Introduction to Paging Systems	42	0974694371	$14.99	2004
BK7769436	Introduction to Private Land Mobile Radio	52	0974694363	$14.99	2004
BK7727878	Introduction to Satellite Systems	72	0974278785	$14.99	2005
BK7781312	Introduction to WCDMA	112	1932813128	$14.99	2004
BK7727879	Introduction to Wireless Systems, 2nd Edition	76	0974278793	$19.99	2006
BK7781337	Mobile Systems	468	1932813373	$39.99	2007
BK7780534	Wireless Systems	536	0972805346	$34.99	2004
BK7781303	Wireless Technology Basics	50	1932813039	$12.99	2004
Optical					
BK7781365	Optical Dictionary	712	1932813659	$39.99	2007
BK7781386	Fiber Optic Basics	316	1932813861	$34.99	2006
BK7781329	Introduction to Optical Communication	132	1932813292	$14.99	2006
Marketing					
BK7781323	Web Marketing Dictionary	688	1932813233	$39.99	2007
BK7781318	Introduction to eMail Marketing	88	1932813187	$19.99	2007
BK7781322	Introduction to Internet AdWord Marketing	92	1932813225	$19.99	2007
BK7781320	Introduction to Internet Affiliate Marketing	88	1932813209	$19.99	2007
BK7781317	Introduction to Internet Marketing	104	1932813292	$19.99	2006
BK7781317	Introduction to Search Engine Optimization (SEO)	84	1932813179	$19.99	2007
Programming					
BK7781300	Introduction to xHTML:	58	1932813004	$14.99	2004
BK7727875	Wireless Markup Language (WML)	287	0974278750	$34.99	2003
Datacom					
BK7781331	Datacom Basics	324	1932813314	$39.99	2007
BK7781355	Introduction to Data Multicasting	104	1932813551	$19.99	
BK7727873	Introduction to Data Networks, 2nd Edition	64	0974278734	$19.99	2006
Cable Television					
BK7781371	Cable Television Dictionary	628	1932813713	$39.99	2007
BK7780536	Introduction to Cable Television, 2nd Edition	96	0972805362	$19.99	2006
BK7781380	Introduction to DOCSIS	104	1932813802	$19.99	2007
Business					
BK7781368	Career Coach	92	1932813683	$14.99	2006
BK7781359	How to Get Private Business Loans	56	1932813594	$14.99	2005
BK7781369	Sales Representative Agreements	96	1932813691	$19.99	2007
BK7781364	Efficient Selling	156	1932813640	$24.99	2007

For a complete list please visit
www.AlthosBooks.com

Printed in the United States
118195LV00002B/58/P